ADOPTING A DIGITAL WORKFORCE

Leadership for successful business transformation and happier people

MAURICE DUBEY and JOHN BURNS

Designed by Shaun Jury
Printed and bound by Bluestar, Wellington
First published in 2020 through Point Publishing Limited

ISBN 978-0-473-55288-6 (paperback)
ISBN 978-0-473-55290-9 (Kindle)
ISBN 978-0-473-55289-3 (EPUB)

"The robots are coming into the workforce and we are just scratching the surface around how to address the people-issues of making this work in practice. Many organizations struggle with this, which is one of the underlying reasons that automation initiatives fail to scale. *Adopting a Digital Workforce* is one of the first books to explore these issues at a practical level."
George Lawton, technology journalist and contributor, California

"I come across so many companies actively working to improve outcomes. Yet they often lack effective organisational change management. Communication for adoption and proactively dealing with resistance are issues that surface time and again. This book tackles all of that with its focus on a people-centered approach. The book primarily looks at implementing strategic technologies, but there is much here that is just plain good change management."
Dan Skelsey, consultant, trainer, contributing author to the Change Management Body of Knowledge, *and author of* The Effective Change Manager's Handbook

"It is great to see a book that analyses and explains the intricacies of moving to a digital workforce. Companies struggle to comprehend the nuances behind the transformation journey, and *Adopting a Digital Workforce* explores this in detail with examples and practices that strategic leaders of companies should be listening to. This book answers some of those tricky questions you didn't know you had."
Craig Richardson, Chief Information Officer, MMC

CONTENTS

ABOUT THE AUTHORS

Maurice Dubey

Maurice is the Executive Director of Q4 Associates, the digital workforce specialists. He's dedicated the past 25 years to bridging the gap between IT and business, and more than a decade to project and program remediation across businesses in the USA, New Zealand, and Australia.

During his time holding multiple senior-level and executive positions, Maurice has developed an eye for what's critical and what's not. He sees the potential of technology not for technology's sake, but for its potential to bring about lasting, meaningful change for businesses, people, and society.

Maurice has spent years at the coalface helping accelerate the establishment of digital workforces in some of the most significant organizations across Australasia. With a global network and experience, Maurice brings his thought leadership to conversations about the optimal adoption of this technology and the next evolution of work.

In addition to multiple industry-specific qualifications, and being a certified Blue Prism ROM (Robotic Operating Model) Architect, Maurice also holds a postgraduate qualification in International Strategy and Leadership, and a Masters in Management from the University of Auckland, New Zealand. He is also a member of the Institute of Directors in New Zealand.

John Burns

John is a business technology consultant and an accredited trainer with over 25 years of experience in technology, strategy, and governance. He's worked in multiple market verticals across New Zealand, the USA, the UK, Europe, and Australia.

Previously director and majority owner of a US software company, the combination of John's strong commercial acumen and insightful understanding of technology bodes well for the business consultancy and strategy work he undertakes for boards and executive teams. John sits and advises on several boards, and focuses now on the not-for-profit space—giving back to organizations that otherwise cannot access the support they need.

John has extensive experience in strategy and has been responsible for the adoption of digital workforces from the initial stages. This sees him focus on technology as the enabler of better business outcomes rather than a driver in its own right.

His specialist skills, ability to combine leadership with vision, and passion for coaching and strategy have seen him deliver exceptional value for major organizations across the world within the private sector, government and not-for-profit space. He is drawn to helping drive organizations to do better through the alignment of technology and business.

A qualified Agile, Change Management, Better Business Cases, and Prince 2 practitioner with an Honors degree in Software Engineering, John is also a member of the Institute of Directors in New Zealand.

INTRODUCTION

Organizations and change

Organizations, in some variation or another, have been around for centuries. Groups of people who have traded goods, differentiated themselves from their competition, turned a profit, found purpose, and played their part in the wider society and economy.

It's a bit of an understatement to say that the way that businesses and people do all of this has changed over time. Even just looking at relatively recent periods, from the first industrial revolution (following the invention of the steam engine) to the second (which saw us firmly in the age of science and mass production from the late 19th century onwards), our modern society has been transformed by the art of business.

So, too, has society changed with the digital revolution. We've seen the exponential rise of data-processing capability, which began in the middle of the past century with the invention of computing machines, be quickly followed by the proliferation of personal, network, and then cloud computing. Like the earlier industrial revolutions, the changes resulting from these 20th and 21st century technologies—and the impact they have had on the way we live and work—are profound.

The disciplines of business, from development and research to design and manufacturing and from marketing and sales through to fulfillment, evolve as the world learns. We learn to think differently and to cater to new demands, desires, and fears. And while significant innovations change the shape of society and the modes of operating for businesses, the years that follow those great leaps are largely spent refining the prevailing ideas that drive the times. We develop new theories, and adopt and adapt to these. Change, then, feels constant.

Change isn't as simple as just flicking a switch. While (some) organizations can refocus or change direction on a dime, it takes time for the people who make up that organization to change and adapt. Implementing a new system within a business is one thing; changing the societal structures—those tenacious habits, legacy processes, and preferences—that sit around it is another story. It is a much messier and more time-consuming process to change an organization's culture, or to meaningfully sway an individual's attitudes.

Where do we find ourselves now?

The changes are coming thicker and faster than ever before. That's not just consulting hype. New technologies are being introduced almost daily, with the expectation that businesses will keep up or risk falling behind. While the rate of change has been accelerating for many years, it hasn't always involved significant or seismic shifts—and it hasn't been big enough to meaningfully change business models.

Now, as we enter the new age of the digital workforce, entirely new business models—driven by new technologies—are emerging. We've heard many commentators say that the technology itself is the game-changer, but instead we believe it to be the *catalyst*. The outcome is that more and more businesses are rising to meet changing consumer behaviors and societal attitudes.

Fintechs are a prime example. Innovative organizations that provide financial services through (digital) technology, fintechs have many of the more traditional banks running scared. Where the latter have been subject to heavy regulatory changes that make agility difficult, fintechs are starting and launching new services that meet changing consumer demands within a year.

NYC-based property insurance giant Lemonade is one such example. The company makes use of artificial intelligence and chatbots to deliver private insurance policies and handle claims for renters and homeowners. Lemonade's technology entrepreneur founders have taken one of the least trusted sectors of the economy—one with extremely low satisfaction scores—and created a new model that aims to swing incentives back toward consumers, amplify transparency, and promote social good. Their fully automated workforce services an area spanning 22 US states, from California across to Pennsylvania. And they have zero insurance brokers.

They're not the only disruptor. Uber, Airbnb, Netflix, and Tesla have all ruffled the feathers of more traditional industries and transformed the ways we commute, holiday, entertain, and travel, through the use of disruptive technologies. However, what makes Lemonade remarkable is the way that they've not just innovated their offering to the market but have also changed the way the people within the organization work (in their case, a pretty small number of people). They've used technology to drive innovation, and they've done this by unlocking the potential of automation.

So where has all this automation come from? And what do we need to know in order to understand how and why it's been so successful?

Automation as a concept is not new—it's been alive and kicking for centuries in some form or another—but its applications have grown and evolved as we've looked to save labor and improve efficiency and accuracy in our daily tasks. Early last century we saw physical robots in factories perform manual manufacturing tasks with better speed and fewer errors than humans. More recently, we've seen automation move into software as people began writing apps to automate basic tasks and free up their time. In some cases these scripts were bolted onto existing business process management (BPM), for example triggering a request to a person and attaching the PDF of the document they would need to fulfill that request. Since then, we've moved from enhancing existing software solutions to full-blown robotic process automation (RPA), which utilizes a whole new toolset to perform digital tasks. In a nutshell—and a long way away from the physical factory robots of old—RPA utilizes software to interact with existing applications and undertake repetitive processes or admin-heavy jobs within an organization.

We're now seeing the development and deployment of greater and greater digital workforce capability. So what does this really mean? Well, a digital workforce carries out tasks through the use of software robots that run either autonomously or as prompted by people. Similar to a pianola—a piano that plays music by itself—the robots utilize various software solutions to perform predefined activities. While there are instances of fully digital workforces like that of Lemonade, more commonly we see the creation of *augmented* digital workforces within organizations—i.e. automation running in conjunction with people.

With augmentation, the software performs specific work activities to a predetermined point. If some level of cognitive processing or judgment is required, such as for particularly complex transactions, the software escalates (refers) the task to a person for processing. Humans still deal with the exceptions and handle data that lies outside of the norm. Augmented digital workforces are a real boon: they're reusable and relatively cheap, and the robots don't need to do everything; all the easy things are handled by robots, while the complex and more cognitively heavy functions get passed to humans, where our problem-solving abilities shine.

Digital workforce solutions can encompass the use of multiple technologies, from optical character recognition (OCR), which extracts text from a document, to artificial intelligence (AI), which translates unstructured data into structured data or determines the meaning of text. That said, the majority of digital workforce solutions rely on RPA as the foundation or backbone that other technologies branch off to perform more specific functions. This is what this book focuses on.

The benefits of digital automation

Unlocking the benefits of digital automation sounds wonderful. Indeed, when implemented well, digital workforces are succeeding in reducing organizations' ineffective bloat and improving their agility. While initially the focus or primary perceived benefit is often cost savings, organizations are now beginning to understand and unlock vastly more business value.

For starters, there's the opportunity to improve customer experience, which was a key driver for Lemonade's establishment. Robots can do repetitive activities more quickly, more accurately, and more tirelessly than humans can. The potential for human error is removed, and robots can cope with greater speed—especially given that they work 24 hours a day and never take holidays, tea breaks, or sick leave.

Digital workforces can also be adapted very quickly to changing demands, requirements, and legislation. Businesses in highly regulated industries, like financial services, are finding that with its increased accuracy, automation provides an effective way to apply consistency to challenges around compliance.

Because humans get on board with new ways of doing things at different rates, a human-centered approach to change can be messy. We all know how challenging it can be to adapt the way a human workforce does something in response to a regulatory change, especially when you've got a team of 50 performing that function. You hold a training session—but a couple of people in the team are away sick that day, so you hold a second training session to include them, too. Slowly, the team members are starting to get it, but they're lacking consistency and you're possibly exposed to compliance risk. Six months down the track, an employee returns from long-term leave;

they start doing things the old way until someone realizes and runs through the training again with them.

RPA can simplify and synchronize a change. You'll no longer need to train people up to perform a function differently; instead, you will pre-configure the change in all of your robots and simply turn them on when the change needs to go live. Robots don't go through the change curve that people do; they don't experience the associated emotional reactions.

Alongside organizational agility are the benefits associated with scalability. Not only do digital workforces adapt quickly to change, but they can also quickly cope with more throughput. Let's say you do a marketing campaign that goes incredibly well. A department that has digital workforce capability may not need to recruit and induct or train more human resources—they can simply deploy more robots for additional processing power as they scale delivery. They're up-scaling with a fraction of the resources and time that they would have needed otherwise.

Those organizations that are doing automation well have set the stage for increased AI in their businesses, having laid the groundwork to gradually educate and configure their robots to do more and more work. However, automation is not about robots taking over the world or, worse, ousting us all from our jobs—quite the opposite. Most digital workforce capability focuses on freeing up human capital to perform the creative and high-impact activities that will move organizations forward.

A McKinsey report predicts that somewhere between 75 million and 375 million job roles will be displaced by automation by 2030.[1] However, Leslie Willcocks, Professor of

Work, Technology and Globalization at the London School of Economics, attests that "in the longer term, RPA means people will have more interesting work."[2] In his view, automation will relieve some of the pressure on workloads that an exponential increase in data, regulation, and bureaucracy has brought—pressure that has seen us, by necessity, making jobs increasingly uninteresting and de-skilled over the past 130 years.

This sounds wonderful, but unfortunately the reality can be quite different. While some organizations are building digital workforce capability successfully, most aren't. In 2019, research by Professor Willcocks into organizations attempting to implement RPA programs globally found that around 20 percent of them are doing it well, scaling and achieving the desired return on investment (ROI).[3] A further 35 percent are successfully implementing it and starting to see positive gains. Unfortunately, another 35 percent are struggling to get any positive results, and the final 10 percent have stalled or are shutting down their programs.

Where to from here?

The thinking and technology surrounding RPA is still in its infancy; it's evolving as it goes. Serious implementations of RPA have only really been operating since 2010, so while the technology, governance, and delivery approaches are continuously maturing, they are remarkably young—and, in many ways, juvenile compared with software development practices.

Adding to an absence of positive examples and use cases, the relationship between technology and people has to change.

This is difficult. A manager in a large corporate may be used to having 50 people in their team. Now they have two, plus a host of robots. While traditional models of remuneration remain linked to how many heads you're managing, this change is unlikely to be met positively. There are ways to help people feel as though they're still managing a workforce, and a digital workforce is more like a human team than a software project in terms of management—as you will see in our later sections. Implementing a digital workforce is not ideally a job for IT!

It's also worth understanding that the process of implementing a digital workforce requires seismic changes for people, which we know take time. Today, it's no longer simply a case of businesses having to evolve faster to meet the rate of change. They are now having to change gears and rethink how their business operates, what their model needs to be, how to stay relevant and sustainable, and how they need to treat their incumbent workforce through these changing times.

It's easy to understand why many organizations are struggling, and why so few are obtaining the full benefits of doing digital workforce adoption really well.

For some, it's a case of "bad but recoverable": an organization is struggling to scale and get processes working, or they're not yet achieving a positive ROI. This isn't a good scenario for the teams that aren't delivering or aren't successfully demonstrating that their efforts have been worth it. They are likely tarnishing RPA's reputation and turning the executive team or the board off the whole concept. This is particularly bad when their competitors are reaping the substantial benefits that can be achieved. However, these scenarios are recoverable.

Even when things haven't been set up as efficiently as they could be, there's the ability to go back and look at the overall program, refine the methodologies, and get the right skillsets involved—if someone is willing to wade through the chaos.

For other implementations, we might see programs that have the wrong message, the wrong (or no) change management, and completely the wrong culture surrounding them. The programs are facing constant push-back within the organization and they often get shelved indefinitely—or at least until a competitor's success in the space spurs the organization to take another look at tackling it. However, even in the latter scenario, we believe that things are recoverable and any rescue efforts will be worth it.

The most important consideration is strategy. Tactically, automation can be used for cost savings, but if you consider it as a broader strategic tool, you will undoubtedly get more out of it. We've seen time and again the incredible benefits that successful digital workforces can unlock, but also how significant the impact of poor implementation and low adoption can be. And that's one of our primary drivers behind the decision to write this book.

Actually, our reasons for writing this book are two-fold. First, there is the huge and evident gap in industry knowledge. We've pointed to the statistics already. We have witnessed and heard stories about too many of the same mistakes being repeated again and again, by companies all the way across the globe. Truth be told, there's no single aspect of RPA that is particularly difficult from a technology perspective. But for the establishment of a digital workforce to be successful within any organization, there are many (far-reaching) aspects that need to be considered and given conscious action. And that's simply not happening sufficiently. Closer to home, we've seen too many supplier and industry presentations done by people who are taking a view that is much too narrow. In our opinion, they are simply not focusing on the wider areas that ensure greater success.

This book aims to change that, with its specific focus on the key business activities and mindsets required to increase your likelihood of success when implementing RPA. It will also enable greater success with any of the additional technologies you may use to enrich your digital workforce.

It does this because it considers things strategically.

We hear you asking what makes us qualified to comment—and how do we get to talk with authority about the key ingredients for successful RPA establishment? For starters, we've worked with some of the biggest organizations in Australasia and the world, including telcos, multinational banks, regional health boards, and logistics businesses. We've worked with smaller businesses, too—local lawyers, tailors, and not-for-profits. Between us we have half a century of experience in helping organizations realize their business goals by bridging the divide between technology and the rest of a business. Whether real or imagined, this divide is something we don't think should exist at all, and it's a gap that we hope is diminishing. That's some of the reason for us writing this book.

But this book is not just a reflection of us. We have been lucky to be backed by a team of experts and to work with some incredibly forward-thinking companies. The stories within these pages are examples we have witnessed, iterated, and learned from in our years in business. It is designed to be an interactive read. You can highlight sections, refer back to things, and add your own notes and learnings. We hope you will—so feel free to scribble!

We have set out to propose a framework for successfully adopting digital workforce capability within an organization, taking a strategic and holistic view to creating sustainable change and realizing tangible business value. As well as being technology- and tool-agnostic, this book is also divisionally agnostic. Different sections will be relevant to people from different levels of your organization.

We start by exploring the opportunities and implications in the era of the digital workforce, and explore the potential that exists for true transformation. Our intention here is to help you understand the potential while also sensibly sorting through the hype to find reality. We address the keystone principles and a few common misconceptions, before outlining the strategic considerations for success and the vital importance of sound change management. Finally, we dive into the selection, delivery, and running of your RPA phases in more detail with a framework that brings the different threads together into the best and most holistic approach.

This book is not a technical or step-by-step guide. It is a commercially savvy "how-to" at the strategic thinking level. It will provide the context necessary to see real, lasting digital workforce success. While we will cover some basic RPA context—what it is and is not, the benefits, and so on—we won't be rehashing what screeds of vendor-specific content already says.

Instead, we will share and recommend the activities that we have researched, implemented, and witnessed that have been beneficial to the success of RPA in organizations. From strategic considerations to advice for influencing others throughout the organization, you'll learn how to deliver value to the business quickly through RPA while successfully managing the people in your organization throughout the process.

Earlier we mentioned the rapid rate of change, which often sees people and organizations out of alignment. This brings us to the second reason for writing this book: Quite simply, we have seen too much evidence of a lack of attention being paid to the *human element* of implementing RPA. Most organizations are not managing their internal change management very well, nor looking out for their people in the process. We do not believe that this is necessarily happening unethically. More likely it is that in failing to bring together all the requisite parts for successful RPA, people are being overlooked. There is irony here, as the human element is a critical ingredient for RPA success. After all, businesses are not made up of processes—they're made up of people, and these people need to be a primary consideration. There is a way forward that works, and this book will help you find it.

If the previous shifts in the world of work are anything to go by, digital workforces will continue to evolve and change. From the relatively basic and repetitive tasks that they are able to do in organizations, they will become more sophisticated in their capability through the adoption of other technologies such as AI.

We truly believe that the adoption of digital workforces

within organizations today can have a lasting and positive impact on the whole future of work—benefitting businesses, people and society itself. This isn't just about business transformation; it's about the global transformation of work.

This whole area is so fast-moving that organizations are trying new approaches and extending their techniques and understanding every day. We understand that even as we are writing this book to some extent it is already out of date. To help with this, we have created the following website where you are welcome to submit feedback and your own learnings. We will also commit to continuing to share updated information as it is learned.

Change is here and it is constant. It needs to be managed better, for all of our sakes.

www.DCAF-book.org

SECTION 1

Transformation: The real opportunity of a digital workforce

We get excited about digital workforces. Not just because what they are capable of doing is amazing, but also because of how they disrupt more traditional ways of doing things—and for the better.

However, before we dive in to our favorite topic, we want to quickly cover off some of the terminology we use. Most vendors and consulting companies try to use their own terms, since this helps to generate a point of differentiation that is useful in their marketing. However, the net result of this is myriad terms that are ultimately confusing rather than helpful—especially when you consider that new terms and descriptions are constantly popping up. Sometimes it can feel like this is happening daily!

We have instituted a few key terms around digital workforces that we believe are more self-explanatory than most. These terms will help people understand where they fit into the wider context of a digital workforce while also helping to promote the successful adoption of this new workforce. For readability and to avoid confusion, we'll endeavor to stick to a consistent set of terminology throughout this book. Here we offer a simple picture of the most common terms you'll come across.

Organizational workforce/augmented workforce

As illustrated in the workforce model in Figure 1, your **organizational workforce** is made up of both your existing human workforce and a digital workforce. Ideally the digital workforce will be working *beside* your existing human workforce—in other words, the human workforce is **augmented** by the digital workforce, as opposed to being replaced by it. With that in mind, both workforces need to be established, educated, trained, and managed to encourage them to work together.

A **digital workforce** is a generic name for a collaboration of technologies that perform business functions. Usually the core technology in a digital workforce is **robotic process automation (RPA)**, but this is not always the case. Other terms used for a digital workforce that you might come across include *connected RPA* and *intelligent automation*. And there are plenty of others.

The key point to remember is:

> A digital workforce is the overarching term for the technologies that work in collaboration with each other and work together with your human workforce.

Digital colleague

Beyond the overarching term *digital workforce*, things can start to get a little gray since we often need to make a distinction between the business's conceptual view and the physical technology set-up and software licensing. This can be made even

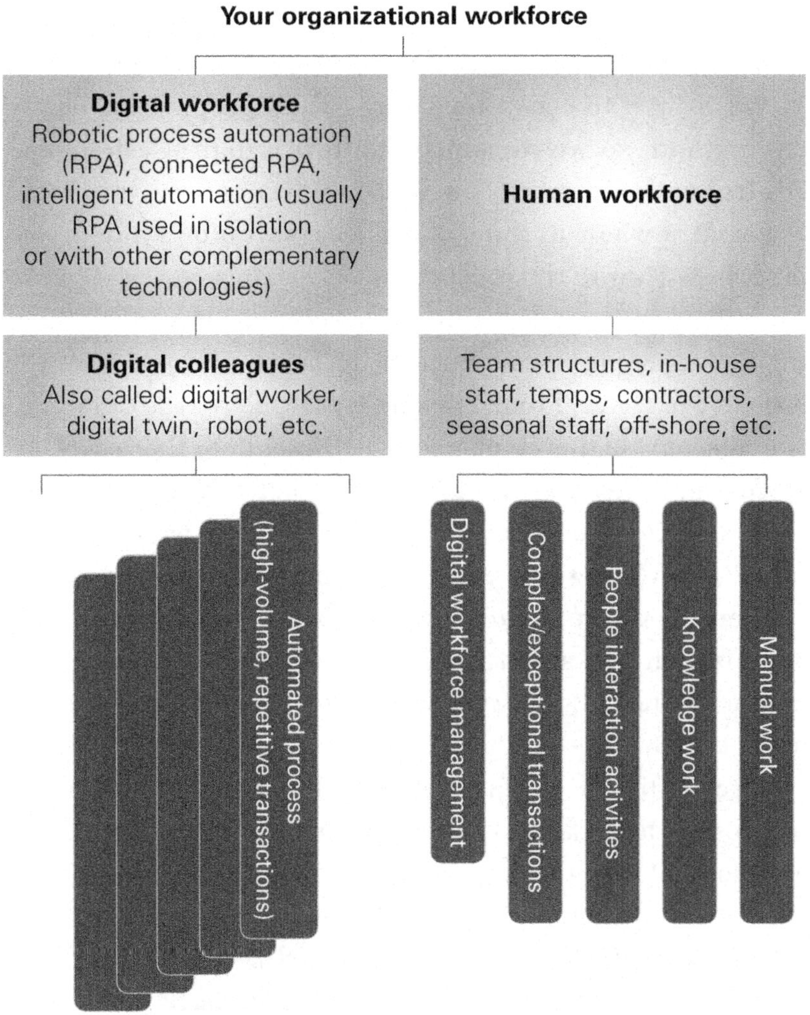

Figure 1: Workforce model—digital workforce operating alongside the human workforce

more complicated by the various ways in which organizations choose to set up their digital workforce technology platforms.

We prefer to focus on things that people can easily understand, so we've coined the term "digital colleague." **Digital colleagues** are simply non-physical (or non-human) workmates who run some of their organization's automated processes. They're the robots themselves. In some cases these automated processes are run with no human interaction; other times they require some form of initiation or input from the human workforce. From a business perspective the difference is irrelevant, so there's no need to differentiate these aspects.

Digital colleagues can also be—and often are—known as *robots*, *digital workers*, or *digital twins*, and we have even heard them being referred to as "robos." We feel that these workers should be treated like someone a person within your organization works alongside in a collaborative and synergistic manner, hence "digital colleagues." We'll explore this further in Section 3.

It seems like an obvious statement to say that words matter, but research shows that we respond to words at a visceral, autonomic level; the way we talk about things affects how we feel about them. During his time as a doctoral candidate at the University of Michigan's Department of Psychology, David Hauser partnered with his colleague Norbert Schwarz, Provost Professor of Psychology at the University of South Carolina, to conduct five experiments aimed at uncovering how the everyday context of a word sways people's judgment and decision-making.

"Some words have a meaning to them that we don't often

think about but yet still affects us, which has applications to persuasion, social influence and bias in our judgments and decisions," Hauser explained.[4]

How does this affect a digital workforce? Let's start with a look at the term "worker"—as in *digital worker*. A *worker* is typically considered an employee or a laborer, usually blue-collar, and historically the term could have been substituted for the words *peasant*, *proletarian*, *slave*, or *serf*.

Now let's take a look at the idea of a "colleague." Suddenly our minds are awash with words like *teammate*, *ally*, or *partner*. We think of a companion or comrade, even a buddy. A digital colleague is therefore someone (or something) that your people work alongside and collaborate with—in much the same way as they would with a human colleague. This understanding is what we need in order to maximize the benefits of a digital workforce within an organization. If we continue to think of robots as just pieces of technology, our digital workforce programs are destined to fail; so we need a mindset shift, and for that we need a name change.

Digital colleagues *help* your people—whether by providing customers with more up-to-date information or faster turn-arounds for a better experience, or by taking the load off already overstretched teams. Organizations are still, and will remain, composed of *people*. Digital colleagues merely represent the next great chapter in human–machine collaboration, if we can evolve our mindsets to see them as such.

Automated process

For us, an **automated process** refers to a configured solution that runs a business activity or process. It will usually interact with other technologies and applications within your organization.

Where you are using RPA as the primary technology for your digital workforce, your configured automated process is run by software. It's really the automated process that is interacting with your human workforce, but it's conceptually easier to think of the people in your business as interacting with digital colleagues.

Introducing potentially transformative (or disruptive) technology

While this book deals with the successful implementation of a digital workforce, robotic process automation (RPA) is just one of many technologies that can assist with business-wide transformation when adopted successfully. Much of what we are saying, and the level of thinking we are advocating for, is relevant for bringing to life any potentially disruptive technology within an organization. It applies whether you're just starting off on your journey to implementing a digital workforce or similar technology, or are already in the throes of doing so. Even if you are some way through establishing a program, we'll highlight strategies throughout the book that you can circle back to look at.

Without a doubt, digital workforces, RPA, and technologies from big data to artificial intelligence (AI) and cognitive and machine learning have the potential to be transformational and to have a profound impact on the way companies operate. They can be used to disrupt the status quo, gain market share, leapfrog the competition, and drive consumer (or even societal) change. However, they don't always lead to *transformation*.

Take big data as an example. The amount of information available to organizations today is transformational—but only if organizations take the opportunity to translate it into insights, and then use these to streamline operations or position themselves ahead of the competition. If a business is not set up to utilize the masses of information they collect, then it all becomes just another data store taking up space somewhere.

> It is mindset and approach to implementation—more than the technology itself—that makes the difference as to whether or not a business transforms.

There is no doubt that new technologies are opening up new opportunities for businesses, but the technology itself is not the transformation—it's the *enabler*. Also bear in mind that some change is more about refinement than about the true transformation that drives a revolution.

Understanding this lays the foundation for success when it comes to implementing a digital workforce and realizing its full benefits in your business. True transformation always has been, and always will be, about people, paradigms, and culture.

The first question: Why?

The starting point for implementing transformational technology should always be the goals of the organization, followed by the selection of new technologies to help achieve those goals. What is your business looking to achieve? Why are you doing it? Without technology selection being driven by organizational goals, it is easy to proceed down the path of short-term tactical decisions without any strategic direction.

In Lewis Carroll's *Alice in Wonderland*, Alice comes to a wood and sees the Cheshire Cat sitting in a tree. She asks it which way she should go from there, to which the cat replies, "That depends a good deal on where you want to get to." Alice answers, "I don't much care where." "Then," says the cat, "it doesn't matter which way you go." The message is simple: you need to know where you're going in order to make a decision about, or plan for, how to get there.

While we might like to believe that businesses are making rational decisions around technology selection, we know that many purchasing decisions are in fact based on emotions rather than logic. After all, emotions influence our attitudes and judgments, which in turn influence the decisions we make. PR expert, motivational speaker, and author Michael Levine says that rationality (including practicality and objectivity) only represents about 20 percent of human decision-making.[5] For this reason, we must consider other drivers.

It may seem simplistic to urge organizations to slow down and consider their reasons before launching into a new technology solution, but the emergence of transformative technologies creates an opportunity *only* when those technologies are successfully adopted and leveraged to deliver tangible value to a business and its people. We're still seeing far too many instances of digital workforces becoming just another technology project for the sake of technology. Worse still, we're encountering islands of technology within a business along with focused areas of change that don't cover anywhere close to the whole business.

The use of a digital workforce can support a huge range of business goals. However, it is critical to define the key reasons why you are implementing one, and to provide clear ownership and a clear linkage to the goals and strategy. Otherwise you'll be like Alice in the wood, going nowhere in a hurry.

Here are just some of the reasons why a business might start down the digital workforce route:

- to increase market share through improved customer service and innovative services
- to become an employer of choice and attract and retain the best talent
- to garner efficiency gains and deliver greater shareholder value
- to reduce risk and improve compliance
- to increase organizational agility and scalability.

However, when we see businesses looking at starting their journey toward integrating a digital workforce, or when we help them to pause and re-examine what started them on that journey as part of remedial work, we commonly see three categories of decision:

- "seems like a good idea"
- "keeping up with the Joneses"
- strategy-driven.

"Seems like a good idea"

You might have heard positive-use cases describing what digital workforces are doing. You believe—and rightly so—that what your organization does could be enhanced by a digital workforce successfully and tirelessly working behind the scenes. There's something to be said for adopting a new technology early: you get to innovate, lead the way, get a jump on the competition, and stay ahead of the curve. Depending on how you adopt the new technology, you have the opportunity to change the way you do things or exceed your customers' expectations in a way that nobody is yet expecting. This can all create competitive advantage—in theory.

However, things that simply seem like a good idea, without further strategic thought or connection, usually don't get driven from the top down. Unfortunately the "that's a nice, shiny new toy—let's try it!" approach commonly results in short-term tactical solutions that are not linked to the strategy and usually don't have full organizational support (or sometimes even backing at the executive level). Especially if adoption has started as someone's "pet project." The person driving automation in this case might want to do it for the learning opportunity, or simply because the technology exists, rather than having thought through the benefits for the organization. There may be some wins, at least initially. However, in this scenario there may not be any real or meaningful drivers for the change, which makes it exceedingly difficult from the perspective of change management. And without the accompanying change management, it's unlikely that you will have a significant impact on changing the way the organization does things. An uncoordinated approach to change can herald disaster.

We commented in our introduction that people are resistant and slow to change, and this is part of the reason why the "seems like a good idea" approach falls down. In software development this approach used to be referred to as "skunk works," or "doing things under the covers." It might get the support of a couple of people, but it usually doesn't manage to generate any sense of urgency or effectively "sell" the change to any of the other people throughout the organization who are affected.

"Keeping up with the Joneses" (or keeping pace with the competition)

Another reason that you might be looking at implementing a digital workforce is because you've seen others in your industry start adopting one successfully, and you worry that they might have an edge over you.

There's some merit to this logic. It is often said that if the rate of change and innovation within your business doesn't meet or exceed the rate of innovation in the marketplace, then you're already going out of business. And there is truth to this—we need to adapt or be left behind.

It's also true that if the adoption of new technology by

others in your industry is truly transformative, then you'll see a ripple of changes that you'll have to adapt to, perhaps higher expectations for customer service, speed of delivery, or similar. These are all reasons to jump on the bandwagon.

However, there can be some distinct drawbacks if you're simply "keeping up the with Joneses" and not approaching automation strategically. This keeping-up approach is distinct from the previous approach ("seems like a good idea") in that there appear to be some business drivers involved. However, undue weight is perhaps given to discussions of what's going on *externally* rather than within the organization. Of course you need to consider and respond to what's happening in your market, but any changes also need to be driven by internal consideration such as your strategy, and where your business wants or needs to go.

The word *transformation* is often bandied about as part of this approach, but that's more likely to be because it is what's expected as part of retaining or regaining market share. An overwhelmingly external focus means that most organizations with this as their driver fail to implement a digital workforce, or other technology, wholly successfully within their business. The approach usually ends up with a short-term tactical view fueled by a state of urgency to get the technology change implemented—but without the long-term adoption of a new way of working, which is where most potential benefits lie. For the latter, the requisite care is needed to bring people in the organization along on the journey.

It's also worth considering that you don't necessarily always have to be one of the trailblazers with new technology. There is merit in adopting technology that others have tried, taking on their learnings and then finding a way to do things better. For example, consider the uptake of manufacturing advances in the automotive industry. The UK used to be a powerhouse when it came to manufacturing motorbikes; other countries, like Japan, adopted the technology, improved on it, and further transformed the industry. Now Japan is a dominant force in the creation of motorbikes. The same could be said for the car industry. Using some of the changes that are happening, and applying them for your own strategic advantage, can help to give your company an edge and stop the racing that leads to sub-par implementation.

Strategy-driven: Where does this business need to go?

The last category of decision-making is a justification driven by—and aligned to—a strategic need. This may seem like a subtle shift, but it has a huge ripple effect. This approach turns a technology implementation into a wider, business-driven initiative, and allows the business to realize the true benefits of the technology.

The strategy-driven approach is all about transformation, which we believe is why a digital workforce journey should begin here. When we talk about transformation, we mean rethinking how your business operates. Transformation is not about moving from state A to state B—it's about changing from one paradigm to another with all of the people within the business on board. It involves reshaping the business to deliver products or services in a new way, which in the case of a digital workforce will utilize a new technology. More importantly, transformation is indicative of a shift in people's *thinking*. Transformation delivers the same benefits that the other two approaches might achieve—but also many wider and more significant ones, both internally and externally.

Wherever you are on your implementation journey, it is most important to understand why the decision to implement a new technology has been made currently, and which way you would like to be going. The questions then become:

- How do we wish to harness this new technology?
- How willing are we to adapt?
- To what level do we want to derive business benefits?

A purely technology-driven approach misses the ability to leverage benefits for the whole business that a business-driven approach would enable.

Lastly, as transformation provides a clear link to the organization's strategy, it also offers a compelling justification and succinct explanation for why the rest of the business should get behind the initiative. As a senior executive, imagine being confronted in the elevator by an employee asking "Why are we getting robots to take our jobs away?" A clear and authentic response that would help the employee understand would go something like this: "Because we need to increase our market share due to international competition. Our digital colleagues will help by being able to provide our customers with more up-to-date information and a faster turnaround of services, while at the same time enabling us to help to take the load off our already overstretched teams." This is far better than being caught off-guard with no sensible explanation to give.

Taking off toward transformation

Let's consider an example in action: that of the automated, kiosk-based airline check-in process. If you've flown anytime in the past decade, you'll know that the days of queuing for the two or three check-in booths staffed by airline personnel are long gone. The availability of technology has driven a process change—travelers are now fast-tracked through check-in with automated facial recognition and passports with electronic chips. Our experience as airline customers is extremely different, and the effect within the business of these changes is significant.

The first—or most loudly signaled—benefit of automation and digital workforces is a reduction in staffing costs, or potentially a shift in the focus of those staff to higher-value customers or other higher-value work. The focus of these people might have been moved to the safety and comfort of people on the plane, as opposed to checking them in. You will still see people in uniform floating around the airport check-in area, but these staff step in to assist only when the process isn't going to plan or a customer needs human assistance. The airline staff help customers interact with the technology, and they become the exception handlers rather than the main point of contact.

This is an example of an augmented workforce: people working alongside automation. And if the airline has done its implementation well, those staff will have had the upskilling required to deal with the more-complicated situations that haven't been easy to automate through the kiosks.

Operationally, it's possible that in dissolving the bottleneck at check-in, the business has simply transferred the chokepoint to another area of the airport. So the business needs to broaden its thinking beyond the immediately affected process—check-in—to streamline the entire experience for its customers, or at least minimize the possibility of simply moving the challenges downstream. The physical space planning and logistics surrounding how the airline works with its other partners and suppliers also potentially needs to be reconsidered. It's likely that more floor space is needed for multiple kiosks, while less is needed for counter-based activity. How can that configuration be optimized?

Then, of course, there's us, the customers. If implemented well, automated check-in technology has the potential to transform the way airlines think about customer service. On the day we're traveling, our interactions with, and experience of, the company begin even before check-in. So it follows that the automation has to deliver a service that is reliable, trustworthy, easy to use. And that's just for starters. Beyond that, those companies that are doing it well will be considering what customer service looks like, and what our expectations are, as customers, in this new, more automated world. That will be the mark of true transformation; as we're starting to see, it all depends on the level and degree of adoption.

What transformation is not

Sometimes a simple way to understand what constitutes a true transformation is to understand what transformation is *not*. Many organizations believe that they simply need to go from current state A to future state B, and—bingo!—they've transformed. They believe that they're following a robust transformation approach to improve how the business works, but instead they are simply undertaking a digital implementation that determines their technology strategy and might change the way a subset of the organization will work. Still other businesses come up with a vision that centers around innovation, then jump to upgrade or replace core systems, restructure, or create some new processes. This also isn't transformation.

A classic example of what transformation is not is the trend of moving to cloud computing. Ten years ago, a business might have run all their systems through internal data centers. Now they're shifting—partly, or even wholly—toward cloud-based solutions that deliver the same (or basically similar) business functionality. However, this is a technology implementation, and is not transformational in its own right. Transformation would require the organization to set itself up to leverage the wider benefits of having cloud-based systems—such as completely changing the way the organization is able to work, or responding to employees' increased desire to work from home or with increased mobility. Transformation here could also enable improved agility by focusing IT resources on business needs rather than on keeping servers running.

There are times when moving to the cloud is part of a wider strategic direction for the organization to support the later benefits, and this is where this shift can be transformational—but it isn't necessarily based on the properties of the technology itself. As with any successful strategy, the end goal needs to be clear. Alice needs to know where she's going.

The same goes for the implementation of a digital workforce. If a process can be automated using robots, does this transform the business? The obvious answer is no. If you want to fully leverage a digital workforce, you need to look at your wider operating model—specifically, your *workforce operating model*—and consider what your workforce could look like in the future. Having done this, it's then just a matter of finding the best path to get there by considering how your people need to adopt the change and evolve your culture.

Applying this level of thinking means that you are no longer simply going from state A to state B, but are starting to affect a paradigm shift within your organization. This is what sees you moving from your current way of operating to a new, different, and revolutionary way of operating.

A transformational change requires a shift in mindset, attitude, and culture.

How to view the differences

We have created a lens for looking at a business that focuses on three options for change. Organizations can:

- **refine** the way they do things; or
- **reinvent** themselves; or
- truly **transform**.

It all comes down to the mindset around implementation and how the business adopts and brings to life a new technology.

To help explain the distinct differences between **refinement**, **reinvention**, and **transformation**, we have come up with a simplified view of an organization that we jokingly term the "organizational igloo." Before we dive into the detail, however, it's important to note that there are four key elements of the igloo, and three key principles.

Four key elements

1. The **environment** you operate in. As well as the traditional PESTELE attributes (Political, Economic, Social, Technological, Environmental, Legal, and Ethical), the environment also includes your customers, competition, suppliers, and media.
2. **Operations**, which covers all the day-to-day activities of the business, including—but not limited to—organizational structures, processes, procedures, "running the business" activities, manufacturing, and selling products and services.
3. Your organizational **strategy**—your plan of action that drives the business and defines how you achieve your mission and vision.
4. The **essence** of your organization—your people and, more importantly, the mindset and culture of the

people within your organization. This is what lies at the core.

The relationship between these four elements is illustrated in Figure 2.

Three key principles (or laws of Nature)

While there are many laws of Nature, we see three in particular as being important to remember in relation to this model. (Please bear in mind that we aren't intending to teach you how to suck eggs!)

1. **Change is constant.** All four key elements of the organizational igloo are constantly changing—and need to constantly change to some extent just to stay viable. Adapt or die, as the adage goes!

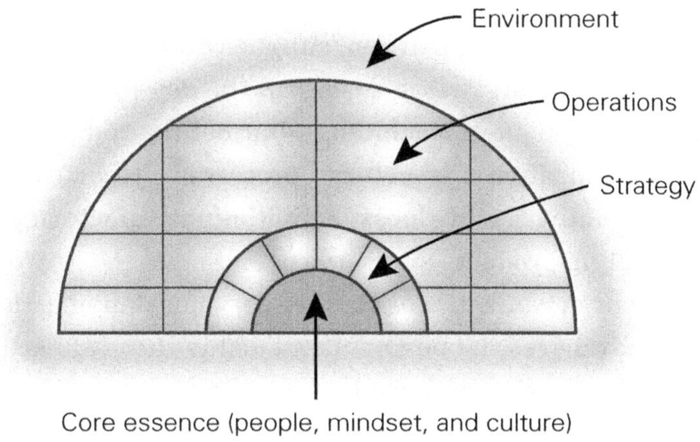

Figure 2: The four key elements of the organizational igloo

2. **Cause and effect**. Change in one area will affect other areas of the igloo. For example, changes in the environment will affect your strategy, people, and operations; changes in operations will affect your people, and so on.
3. **A static object does not influence change.** In our igloo model, only the environment and the core essence (the people) can initiate transformational change.

An easy way to think about continuous change and the law of cause and effect is to visualize a massive snowstorm around our igloo (Figure 3).

Refining and reinventing

We've already covered the concept that there is a wide spectrum of activity that does not constitute transformation—so let's

Figure 3: The snowstorm of continuous change around the organizational igloo

consider this in relation to our igloo. If you encounter cracks in your igloo walls (the operations of your business), simply fixing the cracks—by repairing or changing processes, or amending operations— is not transforming the igloo. Nor is it transforming your organization. Because we're living in a world of continual change, a one-off change to your operations is not transformational; it's par for the course. If a change is introduced without understanding and managing the influence it has on the other areas of the igloo, then it can only—at best—be a refinement. A digital workforce has the potential to be transformational technology, but it won't actually be transformational if it's just having an impact in the operations layer of your igloo and business.

Let's look at an analogy. In the early days of America's Cup yacht racing, the yachts were wooden boats with traditional sails. All of the teams would tweak and refine their yacht designs and configurations based on these materials in the pursuit of marginal benefits and a slight advantage over the other teams. Here's the problem, though: refinement is little more than "prettying up" the processes or, more crudely, "putting lipstick on a pig." There's nothing inherently wrong with this approach, and there may even be some cost savings and efficiency gains as a result. However, an organization is not going to get the transformation benefits of what they're doing because they haven't understood and captured the flow-on advantages to strategy and—even more importantly—to the mindset and culture that's at the core of the organization.

Now, say that there are some changing conditions in your organization's operating environment that are affecting your igloo. You might re-examine your strategy and implement a change.

You can possibly see where we're going here . . . A simple (or even significant) change to strategy, or even the launch of a new product or service, is simply a one-off reinvention. Yes, there are some advantages to reinventing—but reinventing tends to be reactionary to external influences rather than change being driven in a proactive and ongoing manner. Without the change to the mindset and culture of the organization, even reinvention does not provide true transformational benefits.

Back to our America's Cup team. As fiberglass, Kevlar, wind-tunnels, and computer monitoring all evolved, and the rules of the America's Cup permitted the use of such technological advances, each team had to figure out how to use these materials in their designs for optimal advantage—building and configuring their yachts with new technologies. But perhaps more importantly, they also had to learn how to best operate (sail) these new vessels. They reinvented and changed from state A to state B. If they'd simply swapped wood out for fiberglass, this would have been a refinement far more than a reinvention.

Taking it to the level of transformation

Transformation pushes all of these ideas further, and also takes into account the essence of an organization. Transformation comes from focusing on the mindset and culture of the people in an organization in order to continually take a proactive approach and explore new strategies and ways of operating.

Transformation is about having an organization that is

sustainable and evolving. It's a mindset shared by all your people that inspires thinking constantly about opportunities and ways to improve or innovate. It's about continually understanding the needs of the environment, and adjusting to meet these needs with new technology or ways of operating. With all this in mind, we can imagine a continuous cyclical mechanism focused on:

1. **Change.** People driving the change-setting and implementing this based on their understanding of the business, its desired direction, and the environment.
2. **Impact.** Changes being implemented that will have flow-on effects to other areas of the business, producing ripples through the environment, operations, strategy, and your people.
3. **Learning.** Evaluating the wider change and its impacts, to learn and positively influence the mindset of the organization.

Ideally this process is continually repeated, reinforcing the benefits and supporting the mindset of ongoing improvement and evolution at all levels (Figure 4).

For our America's Cup team, it was the skills and attitudes of the designers, boat builders, and crew that needed to transform to make use of the moves to fiberglass or Kevlar—far more than it being the technology itself that transformed them. Their commitment to continually improving all aspects of what they did with these additions made all the difference and took them beyond a one-off reinvention.

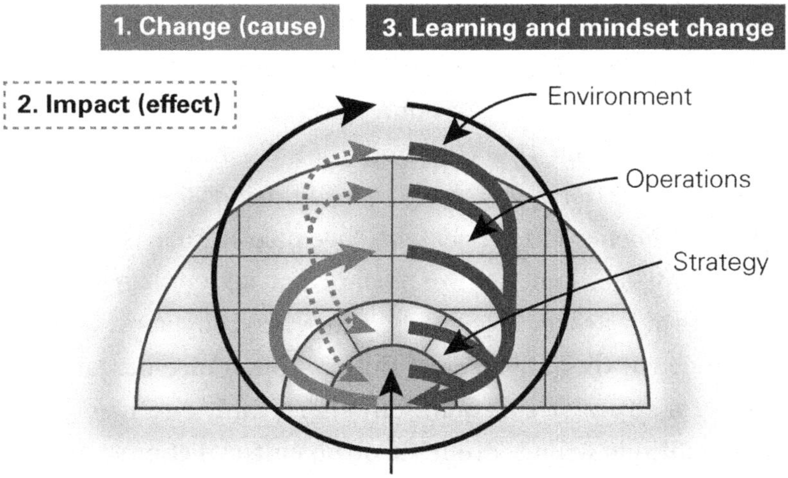

Figure 4: The cyclical mechanism underlying transformation

Each America's Cup team was only permitted to use new technologies when the rules were changed to allow this. All of the teams were therefore on a level playing field, so to speak. In business, however, more often than not there is no dictating body that specifies when organizations can adopt new technologies. This drives ongoing competition and keeps opportunities wide open, but it also means that there is a race between organizations to see who can be the first and the fastest to transform their people's skills and attitudes, and both trial, and learn to maximize the benefits of, new technologies.

Therefore, a robust approach to transformation makes all the difference, especially as we know that you don't just decide you are going to change the culture and mindset of a person, team, or organization. Simply telling someone they need to come into work tomorrow with more customer empathy, or the like, rarely works. As John Narver, Stanley Slater, and Brian Tietje from the University of Washington attest in their article "Creating a market orientation" for the *Journal of Market-Focused Management*, "because culture is a pattern of assumptions based on experience, a new culture or new elements of it are accepted only if they are congruent with the experience of the members of the group and are perceived as offering a superior solution."[6] In short, failing to meet these two conditions will create little or no cultural change.

We'll look further at the change-management aspect of a digital workforce in Section 2. For now, suffice it to say that transformation is about the people in the organization driving change. Narver, Slater, and Tietje's article discusses two approaches to changing culture: a *programmatic approach*, which uses education and changes in an organization to try to implant the desired norm into people; and an *experiential approach*, in which a business learns from its day-to-day efforts, and is therefore continuously developing and adapting. The authors' theory is that both approaches contribute to increased success, but when the programmatic approach is sharply focused on providing a foundation for experiential learning, the combined effect of the two learning strategies is the greatest. They conclude that the two strategies must therefore be tailored and managed as a coordinated joint strategy.

When we implement and manage change, we need to continually improve as we look to understand our environment, and have our people define and take ownership of the required strategy; then we can use the advantages and learnings to influence the organization-wide mindset and culture.

Impact on your strategy

As you focus on driving transformational change with your people, their mindset, and your organization's culture, you will need to evolve your strategy. In the case of a digital workforce, transformation fundamentally affects the *workforce operating model*—the way your workforce runs and delivers value to your customers, especially with and through its workers (both human and digital). It involves a high-level approach to how your business is organized to deliver on your strategy, and it comes with an understanding that the way people do their jobs within the organization will be affected.

Day-to-day we don't really consciously think about our various operating models, or the fact that we even have them. After all, we're often too busy thinking about delivering for our clients and keeping the ship running. But our operating models are still there, working and driving the way that things are done within the business. Alongside the workforce operating model lie other operating models like finance, sales and marketing, and products or services; there may also be others for your specific business. Together, these make up our business model. If we want transformation, we need to think about them—and acknowledge or understand that they are going to change when digital workers are introduced.

Because your workforce operating model is going to become

profoundly different when you successfully implement a digital workforce, it is important to manage those changes proactively. The chances are that your workforce will become *augmented*—made up of both people and digital colleagues. Where managers once had a team of 60 to oversee, they might now have three people and a handful of robots. An employee who once spent their day dealing with simple, repetitive tasks now has electronic "colleagues" taking care of the mundane; this employee now only has the exceptions or challenging cases elevated to them to resolve. Their job likely just got more challenging and engaging, so some level of training or upskilling may well be required to support the transition into their new role.

Specific divisions may have their own strategies around how to structure and source their workforce; finance may have its own strategy for the financial operating model. But it's important for the organization *as a whole* to review and assess these operating models if transformation truly is the goal.

A real-world look at how this might work

Let's say you were to outsource a whole division of your organization to an offshore function. Maybe it's your call center, IT, or manufacturing. In doing so, you are essentially changing your workforce operating model at an organization-wide level. This would need to be evaluated, justified, and approved at board level, and you would need a well-thought-through and well-executed plan to implement the change.

This plan would include, and take into account, things like:

- How will the rest of the business interact with this offshore function going forward?
- What will happen to the existing resources?
- How will the change be communicated?
- How will the initiative be monitored and measured?

When implementing a digital workforce, the same should apply. All the same considerations, approvals, change management, and monitoring activities are required in order for the change to be successful and for the truly transformational benefits of automation to be realized.

It may appear that we are laboring this point—and we are—but that's because we believe that the success of any change depends on it being driven from a *strategic* perspective. For the successful introduction of a digital workforce, it's imperative that the initiative is driven from the strategic workforce operating model, and as part of a wider transformation exercise. This approach sits in contrast to a digital automation initiative that is being driven merely from a refinement approach, at the divisional or process-improvement level within the organization.

Transformation only exists when organizations address the way the people in their business think, and the culture of the organization—from the board to the front-line staff, across all channels and geographies—as well as considering their business model, operating models, and products and services. Everyone within the organization should be tuned to continuously reviewing and predicting the market, technology, and resource needs required to maintain a sustainable and competitive business model.

Undoubtedly this is an additional level of effort to an approach centered on reinvention, but the benefits are significant. We would also argue that having more innovative and curious thinking within an organization, and a culture of continuous learning and development, makes it easier to review operating models and identify new and unique product and service opportunities.

Transformation is not for the faint-hearted

Having said all of the above, we understand that transformation doesn't come easily, and nor is it done lightly. While some organizations are, at their heart, transformational, we understand that for most, transformation is hard. There is often inertia in the way that people do things; we can be slow to pick up new ways and ideas. Plus there will be legacy systems to deal with, and an already deeply entrenched culture. It's a lot to overcome.

We understand that people can't simply dream up new business models, and that many organizations are unwilling to take the risks involved in doing so. We see plenty of organizations that come up with a vision about innovation, or something similar, and then jump in to restructuring or implementing new processes. They're actually only refining—but we also understand why so many organizations tell themselves they are doing transformation when what they are really doing is putting wings on a caterpillar and calling it a butterfly.

A 2014 study by global management consultants Boston Consulting Group (BCG) found that 85 percent of companies claimed to have undertaken a transformation during the past decade. But 75 percent of those transformations failed to improve either short- or long-term business performance.[7]

Virtually everywhere we look, we can see that transformation is high on the strategic agenda; the impetus to innovate is being felt acutely as changes come ever faster to almost every industry. Disrupt, or be disrupted. However, just because it's a business driver, this doesn't mean that these businesses are achieving true transformation, as the BCG study found. KPMG's 2017 Project Management Survey may provide some clues as to why. This study found that only:

- 31 percent of projects deliver on time
- 29 percent of projects deliver on budget
- 33 percent of projects meet business objectives
- 34 percent of projects have satisfied stakeholders (which is unsurprising really, on the basis of the first three points).[8]

A lot of resources are invested into transformation, or even significant attempts to adopt new technology, so we need to sharpen our approach in order to unlock the benefits and deliver business value. We wrote this book to help you understand and achieve this.

Obviously, there are some significant questions related to why so many projects are not deemed to be successful. One common thing we observe—particularly in the case of unmet business objectives and dissatisfaction by the business—is the miscategorization of projects. Transformational technologies are seen as being merely "digital" transformations rather than "business" transformations.

Business vs. digital transformation

Something of a buzzword in the past decade, "digital transformation" is often used to communicate the need to embrace those new technologies that help a business remain competitive and sustainable in today's market. Unfortunately, the name can result in people overlooking the need to drive the digital transformation from a business perspective, and for the whole business to be engaged with the process.

True transformation is about organizational change, and we would go so far as to say that there is actually no such thing as "digital" transformation. That's an idea used by software marketers to sell you the latest update or operating system, and ultimately it results in technology for short-lived market wins. We have never worked on a technology project that doesn't have some level of business implication. In almost every instance, success for those projects has required business leadership and buy-in.

The technology changes inherent in a digital workforce—if done correctly—can have such a profound effect on a business (and its people) that it is hard to see that it is not going to drive a transformational change in most organizations, whether this has been pre-planned or not. Businesses are not merely made up of processes and systems—they are made up of people, and are driven by and for people. So the differences here, and the opportunity to realize the full benefits of any change, lie with them.

Transforming a business depends on transforming individuals.

The hardest change to undertake is cultural, which may be why so many people instead prefer the idea of *just* a digital transformation—anything else appears too hard. However, the only way to radically change or transform a business is to focus first on the people and the way they work, i.e. the organization of operations. And as we've seen, it's not about going from technology state A to technology state B, but about a new way of thinking and operating.

In the article "Leaders focus too much on changing policies, and not enough on changing minds" published in the *Harvard Business Review*, Tony Schwartz outlined the idea that most organizations overlook the internal shift—in what people think and feel—that has to occur to bring a strategy to life.[9] The most effective transformation begins with what's going on inside people, and leaders—with their disproportionate authority and influence—have an immense impact on the collective mindset of an organization and its culture.

Back to the big question

In Section 2 we'll look further at the fundamentals of change management and how they apply to the adoption of new technology and the introduction of a successful digital workforce. For now, you can see that we have arrived back to where this section began—the idea that we need to start with *why?*

Knowing where you are going is the foundation of transformation. Your strategy then is essential to not only planning how to get there, but also inspiring others to come along with you. It ideally creates a clear, concise vision for your transformation. Too often an organization sets out on a journey of refining current processes and procedures without knowing where they are going. This leads to a short-term, blinkered view of the project at hand that doesn't take into account the implications for the rest of the organization.

Perhaps not all organizations need to transform. It's possible to put in a piece of technology, improve the way you operate, and achieve some market wins. For this sort of refinement you may need only a simple definition of the state you want to end up in. However, if this is the approach you take, it will undoubtedly mean missed opportunities around the benefits that can emerge from the successful implementation of a digital workforce rather than a simple RPA implementation.

If you want to implement a digital workforce successfully, and for full benefit, there are some additional things you need to do. Don't worry—you can still dip your toe in by piloting a new technology before it is considered for full implementation; in fact, this is a sensible way to balance the risks. There is nothing wrong with trialing a new technology on a small scale and learning from that trial! One of our clients did just that. After a few false starts with automation via various vendors, a large Australian organization got us on board to help them really get their program under way. We worked with them to refine their "Center of Excellence", and used the first few processes to help them discover what they needed to do in terms of release management, communications, and security checking. In this case it wasn't about testing the technology as a concept—they knew they wanted to adopt automation.

Instead, it was about learning how to do it properly within their organization. They were able to establish their own version of "best practice" that would ultimately work for them.

To set the stage for transformation, you need to understand and deal with the implications of not just the technology but also its impact on the wider business. And you need to do this early in the planning and pilot stages. In later sections we'll look at some of the practical steps around managing change, as well as frameworks to assist with this.

A problem that we constantly see is technology pilots approached from a strictly technology basis. We need to broaden the learnings we are looking for, and the ways we measure and assess a new technology during the pilot stages. It is important to look beyond just technological aspects, and consider:

- What business opportunities are we unlocking with this new technology?
- How will this new technology affect our people?
- How will we need to change our operating models?
- How will we educate people within the business about this change?

As we said at the start of this section, it's important to think about the strategic need in what you are doing. Why do you want to make a transformational change? Are you committed to changing the way you do things? Do you have a compelling vision? Do you understand your current state? Does the change fit with what you do as a business, or what you want to do?

We ask these questions to give you a gauge of your organization's readiness for transformation. Only when you understand these factors can you assess how a digital workforce (or other technology) is going to benefit your business.

> A transformation journey, with a digital workforce or other technology, is about rethinking how you do business. In order to change the culture and attitudes in your business, bringing the "why" into focus is key.

Successfully building digital workforce capability and achieving its full benefits and value for your business depends on big-picture thinking, attention to detail, and focused expertise. A digital workforce works best when you involve the whole organization, not just IT. So, think about what you are doing now. Do you have a transformation program in place, or are you just prettying up the edges? And where are you going?

You don't need to have all the aspects of the new paradigm defined in detail before you can begin the journey to implement a digital workforce. But you *do* want to do some robust "thinking ahead." Too many technology selection and pilot exercises have no thorough evaluation criteria, nor a well-thought-out critique of the results—not from a business impact, a market impact, or a business strategy perspective.

Critical to understanding why you're introducing a digital workforce (or other potentially transformative technology) into a business is the link to a wider strategic objective, and understanding the desired impact this will have on the people, culture, and mindset of the organization. In addition to the obvious cost/benefit opportunities, the implementation of a digital workforce provides an opportunity to transform the organization to deliver more sustainable outcomes. It just makes sense.

SUMMARY

- Understanding the transformational mindset is key to fully leveraging the benefits of a digital workforce, and to helping an organization to adapt, scale, and become more sustainable.
- True transformation requires buy-in and engagement from everyone, right across the organization, as well as subsequent internal and external shifts. The right structures, as well as robust and effective change management, can make all the difference to enabling a digital workforce to mature inside an organization, and grow in impact and reach. Now that is powerfully transformative!

SECTION 2

Implications: What you need to understand as you adopt a digital workforce

We ended the last section with a call to think about your impetus for introducing a digital workforce into your organization. Is it strictly tactical, in the sense that you are hoping to refine a single process (or a small handful of processes) with RPA? Are you seeing it as more of a straight-forward technology project? Or is automation part of your strategic goals, in which case transformation via a full digital workforce is the name of the game?

We are deliberately stressing this point in order to set up the level of thinking that's required as you embark on potentially transformative technology programs, like the implementation of a digital workforce. Where a technology project like a customer relationship management (CRM) system might have a relatively localized impact, a digital workforce is a prime example of where a change in one area of the business affects *all* of the other aspects—so you really do need to consider the wider implications!

If you are intent on transforming, there is some robust thinking that needs to take place up front. We can help to steer your thinking as we unpack the sorts of things to ask yourself. If you are already aware of the need to do some big-picture exploring, then your mindset is likely close to the right place.

Do you believe that you are working to change just one process (or a few)? The plan might be a simple RPA implementation as opposed to the creation of a full digital workforce. But even if you don't believe that your pilot or project is linked to any sort of strategic driver, you still need to broaden your thinking. Because of its far-reaching business implications, the implementation of a digital workforce cannot be a solely technology-driven project. Sure, you may get some benefits and cost savings from the technology. But we recommend that you take a wider view of its implications in order to understand the full scope of the potential benefits.

Understanding the implications of transformation

The amount of work and effort required to understand the challenges of an implementation is directly related to the scope of what you're trying to change. If transformation is a key strategic driver, you'll have to put a lot more upfront effort into understanding what the implications are for your organization. If you're just looking to change or add a process or two, you still need to think about all of the same things, but the implications of those changes might be limited to a part of your organization, and the internal messaging will be quite different.

While the link to strategy is important here, so too is the pace of adoption that you're aiming for. If you're planning for a fairly gradual approach of delivery, learning, and maturing, you will be building up your digital workforce incrementally. This allows you to start generating an outcome, and understanding the benefits and impacts as you go. Contrast this with a "Big Bang" or mass-deployment approach. It's easy to see that the latter (often involving BDUF—big design up front) requires far more of the thinking about implications to be done ahead of commencing the project. These two delivery approaches operate through two different mindsets, which has a bearing on how you can best manage the change.

Another consideration is whether your digital workforce journey will begin with a pilot. A pilot makes sense if you're trying to decide whether or not to use a specific technology, as it allows you to test the technology in one area and see if it works before rolling it out further. If you are running a pilot, you might not put in as much effort beforehand around thinking about the implications of the technology change, as that's the purpose of the pilot—it's intended as a learning exercise! In this case, it's important to build into the pilot the outcomes you are hoping to achieve; namely, learnings from the technology change and a greater understanding of the implications of that change on the organization's culture and mindset.

As we've mentioned before, unfortunately most technology pilots are assessed badly, especially when it comes to change management. We've seen countless examples where a pilot is deemed a success and the technology is scaled within the business simply because the results showed that some cost savings were achieved, or that the technology itself worked and integrated with what it should have. That's important to check, of course, but successful integration should be part of, not the basis of, an entire pilot.

Instead, key to thinking about technology adoption in a transformative way—even if you're taking a very tactical approach to your implementation—is consideration of the wider impacts:

- What did you find out about the potential impacts on culture through the pilot?
- What are the lessons for change management?
- How did people react to the change?
- Did it encourage people to consider other business opportunities that have now opened up?

This section of the book will explore some of the questions you might ask yourself and your organization, and will help you consider the level of adoption you require for your digital workforce. If you're only aiming for a limited rollout, you might not need to consider the impacts of the new technology on the business overall, but instead focus on just some key areas. Conversely, if you are aiming for a full strategic rollout, planning around the implications and how you will best introduce and manage the change will make all the difference. You can see that each of these situations will require you to tailor the level of analysis and planning required.

While the ideal is to use the tools and questions covered here *before* you kick off, they can also help you course-correct or adapt from wherever you are in your journey toward a digital workforce. If you are just at the outset of your automation program now, the key message though, is this: Don't blunder into it! You're wiser to consciously think about which way you're going before you start, and use this section to get a full picture of the implications.

The ripples of transformation

In Section 1, Figure 2 (the igloo) showed us the relationships between the key elements of an organization, and we explored what makes for transformation (as well as what doesn't). What we emphasize now is the degree of interconnection between the four key elements. Refining an operational process will have an impact on every level of your business, right down to your organization's core. Like ripples spreading across the surface of a pool, a change might originate from just one area but the effects will spread. This means that digital workforce adoption by stealth won't work.

Think back to our example of the automated airport check-in process. Automating this process through the use of kiosks had other flow-on effects in the airport: reducing the time required to check in potentially meant an increase in the number of checked-in people all at once, the flow-on of which could be a bottleneck in the customs and border-control queues. In this case, the automation would just move the bottleneck to a downstream (later) process. Taking this even further, online check-ins may also change patterns at Customs. It may make the flow of people less predictable, affecting the way we budget time ahead of our flight or how we utilize different areas of an airport. If you have checked in online the night before, you can turn up at Customs very close to boarding time, so Customs staff need to be prepared for this.

The point we're making here is that the delivery of

automation in one area often affects other areas of the organization. This is why you need robust, high-level thinking, even if your change comes from day-to-day efforts.

Outputs versus outcomes

It's worth remembering the difference between an *output* and an *outcome* in terms of the benefits expected by a business. Your deployment of automation should ideally have an objective for your investment that will be linked back to a strategic goal and will deliver one or more benefits.

Outputs are the deliverables, or products, created from your project. An automated process is likely to be an output of your program. The output will do specific things, in a specified way, taking into account the business's requirements and the implementation of the process requirements. We'll cover more of the actual process—happy paths, exception handling, and the like—in the sections that follow. For now, though, it's worth acknowledging that traditional project-management methodologies can be successfully utilized to support the delivery of outputs. A pilot can also be used to ensure that outputs deliver the required functionality. However, solely focusing on the *functional* requirements in a pilot means missing out on some of the potential beneficial outcomes that could come from your project.

Outcomes, meanwhile, are your outputs in use. You won't achieve the desired benefits of your program without successfully utilizing the outputs to generate the required outcomes. Consider a non-technical example to begin with: the US$1 billion tunnel that forms a significant piece of the roading structure in Auckland, New Zealand's largest city. The tunnel itself was an output of a 30-mile (48 km) project called the Western Ring Road. If no one had used the tunnel, this output would have failed to achieve the positive outcomes intended, namely easing congestion by providing an alternative route. Luckily, the tunnel opened successfully in mid-2017 and has provided significant benefits, reducing traffic times across many routes for both commuters and freight.

The outcomes will enable the delivery of the investment objectives, which should ideally be SMART (Specific, Measurable, Achievable, Relevant, and Timebound). A good investment objective for an automated process could be "no more than 20 percent of employee data updates going to human exception-handling per data upload." Or "100 percent of data accurately processed within 30 minutes of the process running." The execution of the process should therefore enable the benefits that were intended from achieving these investment objectives.

In the digital workforce world, we need to always be thinking about the wider organizational workforce—the interaction between digital and human workers, as well as the impact on the human workforce. Change management is driven by a focus on the wider implications, so it's likely that here the thinking will be around other, sometimes less tangible, measures:

- What other influences did the project have?
- How has it affected culture?
- Are your people considering different career-enhancing opportunities now as a result of the implementation?

- How is remuneration, recognition, and status affected?
- Is there other upskilling or training required anywhere throughout the organization as a result?

Of course, project-management principles are helpful as part of the implementation, but transformation is all about broadening the focus, and integrating project management with change-management frameworks sets you up with the best chance for success. A strictly output-based view of digital automation, without the requisite change management, will likely mean that you will miss the full scope of benefits. It's worth thinking about the outcomes you expect, because that's where the potential positive impact exists for the wider workforce and organization.

In the digital workforce industry we've seen a lot of focus on the technical deployment of automated processes, with the frameworks from most RPA vendors providing guidelines as to the best approaches to successfully deliver these outputs. This is helpful, but we want to take things further. We want to look more widely at the integration of the digital and the human workforce to ensure successful outcomes and the achievement of your expected benefits.

Bringing a change-management mindset

So, how can applying a change-management mindset improve our likelihood of success? And what do we need to be thinking about to aid the implementation of automated processes and improve the adoption and integration of our organizational workforce?

Applying a change-management mindset can help us by:

- providing a better view of the areas of impact for the implemented processes, noting that changes in one area can have unforeseen and often wider areas of impact
- providing a better view of any anticipated resistance
- providing a better understanding of business risks, especially around adoption of the new technology.

McKinsey's 7S model for change management

Impact analysis models abound in the academic world, but McKinsey's 7S model (Figure 5) is a simple framework that seeks to identify all of the aspects of an organization that would be affected if you were undertaking a change. It also re-emphasizes the interconnection between aspects of an organization's makeup, further confirming that a change in one area of the business will affect other parts, too. We have found this to be especially true when introducing a digital workforce.

The model provides a high-level assessment only. Various other processes exist for more detailed impact analysis, including stakeholder analysis (detailed analysis per stakeholder group), measures of impact complexity, and organizational change heat maps. We will not go into these aspects in detail in this book, but it may be worth considering them when going through the implementation planning for the digital workforce in your particular organization.

As you might expect, the change to include digital colleagues in your organization's workforce has a big impact on both the hard and the soft elements in your organization. The hard elements—strategy, structure, and systems—are more easily identified and managed. The soft elements—shared values, skills, staff, and style, which relate more to culture—are often less tangible, but are key when considering digital automation. The soft elements are also where we most commonly see problems. Potential changes to an organization's culture, the skillsets required by the human workforce, and the style of management that may be needed in the transformed organization are often overlooked, so receive special focus in this book.

First, let's take a look at each element in turn.

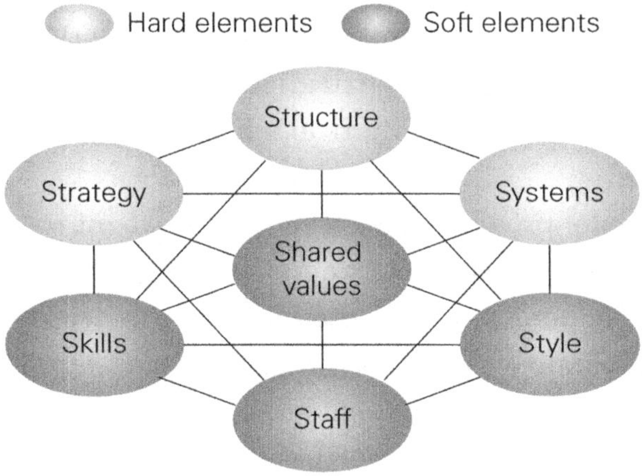

Figure 5: A graphical representation of McKinsey's 7S model as developed by Tom Peters and Robert-Waterman[10]

Soft elements

Shared values, located at the center of this model, mirror what we believe to be the core essence of an organization in our igloo model. This is the heart, mindset, and culture of an organization—the guiding principles around which your organization currently works. You will have come across this idea before; it's often the unspoken, yet pervasive, "way we do things around here."

The *Harvard Business Review* article "Changing company culture requires a movement, not a mandate" by Bryan Walker and Sarah A. Soule describes culture as being "like the wind. It is invisible, yet its effect can be seen and felt."[11] Walker and Soule also make the very valid point that "For organizations seeking to become more adaptive and innovative, culture change is often the most challenging part of the transformation."

We have to agree. Culture change is not simple, and Walker and Soule echo this with their assertion that "culture change can't be achieved through top-down mandate. It lives in the collective hearts and habits of people and their shared perception[s]." So culture plays a decisive role in the success—or otherwise—of transformation, or of how successful your strategic drivers are. Consider how flexible your current culture is. How accepting is it of change? Many company cultures have historically been focused on operational excellence and efficiency, so increased adaptability and new behaviors can sometimes threaten that.

Think back to our airline example again. Where staff used to follow a methodical process to check people in, they now need to be problem-solvers. This can represent a cultural shift. You now need a culture built around the idea that no problem is too difficult; the organization requires a mindset focused intensely on customer experience. At a fundamental level, and certainly from the point of view of the staff, it will feel as though you are changing the way you do business.

A digital workforce may change the way people work in other ways, too. Perhaps the effective implementation of robots means that people are freed up to do creative work that wasn't asked of them before. This might not be what they signed up for! Or imagine a processing hub where a team of people once sat in an office together all day. You might now find a scenario where digital colleagues do the bulk of the (once-manual) processing during the night, while a smaller team of people can work flexible hours, or from home, to check exceptions and manually resolve the more tricky or complicated cases. These people's work hours have changed considerably, as has the social aspect of their work. The ripples of change will undoubtedly affect the team culture.

> Culture cannot be created by someone in authority.
> It relies instead on a process of engendering optimism, trust, commitment, and possibly creativity.

UK-based personal and commercial insurance provider Ageas—whose experience of a digital workforce we'll explore later in this section—has recognized the importance of team culture. They knew they couldn't achieve their process efficiency, waste reduction, and risk-mitigation goals unless everyone in the organization fully understood the benefits of RPA and embraced all it could do for them, the company, and its customers. In a fabulous case study by Blue Prism, a leading RPA platform provider, Ageas established change-management practices to support their digital workforce with the following goals:

- to eliminate worries about the impact on employment
- to actively encourage their employees to welcome process automation
- to create a culture in which employees think of the robots as additional team members who are responsible for doing repetitive, mundane, back-office tasks 24/7/365.

Staff means, quite obviously, the people who make up an organization. Digitizing the workforce has an obvious impact on human employees. Consider how many and what type of staff you'll actually need going forward, and as your digital workforce scales.

Undoubtedly there will be job changes across the organization, or at least in some departments. However, this doesn't always mean redundancies! More likely are changes to the general capabilities of your employees, and an expectation of upskilling as roles change. But if you are considering whether restructuring will be required for your organization, then you'll need to think about this from your people's point of view. Without exceptional communication and transparency, it's common for people to fear losing their jobs, even if they're not involved in the aspects of the organization being automated.

Compared with some of the more common technology implementations, the implementation of a digital workforce tends to have a more primal effect on the human workforce it needs to integrate with. The implementation of a new CRM is unlikely to bring about the same fear of employees being retrenched. Other fears or worries, such as those surrounding upskilling and technological literacy, apply to both.

A well-managed approach to change emphasizes allaying the fears and communicating the opportunities, as opposed to the threats and negatives that people might perceive. While the people component seems obvious, we're glad that this is a specific aspect of the 7S model because, as we've seen, people are often overlooked in the race toward automation.

Skills consist of the general capabilities of your people, and a digital workforce undoubtedly creates shifts here, too. Specialized skills are required to manage the robots, so you need to think about whether this could be an opportunity to upskill your people rather than replace them. Beyond that, additional training is often required for people to fulfill more creative functions within a business as robots take over the more mundane and menial.

Consider our airline example once more. Although staffing numbers reduced, there are still airline staff available around the automated kiosks. Instead of performing a standard process for each customer, they're now handling the exceptions—when there are issues with people following the automated process. This requires a different skillset. These staff would have to be very familiar with the automated process, and how it is supposed to work, in order to be able to troubleshoot for those passengers who seek their assistance. The same goes for the self-checkout area in the supermarket. In these scenarios your staff become more valuable to your entire system, not less.

With a digital workforce, there is often a greater post-implementation reliance on technology. You should consider whether you have an IT-literate workforce, or whether further proactive training in this area needs focus so that your team can stay ahead of technological refinements.

We've already explained the need for effective project management and change-management skills within an organization that is implementing a digital workforce, and usually there's also a need for greater technical acumen among board members and senior executives. It's vital that the leadership is able to discern whether what they are being sold or told about a digital workforce stacks up, and whether this is what's being delivered. Before they're able to judge this, they need to know more about, and understand, what it is they are judging. So, just as most boards would have the financial acumen to understand financial practices and perspectives, they now need technical acumen to deal with technology standards and impacts. Consider whether this exists among your organization's advisors or whether upskilling is required.

There's also the matter of what the new capability landscape of the organization will be once the digital workforce is pulling its weight. What opportunities does the automation unlock in a broader sense, and does the organization have the skills it needs to capitalize on these? Future recruitment and training may well focus on vastly different skillsets and capabilities than what you have historically hunted for, thereby funneling more-diverse talent into your teams.

Style refers to an organization's approach to management. We mentioned previously the idea that managers or leaders, by virtue of their position, hold a disproportionate amount of influence within an organization. If we know that the biggest predictor of a successful transformation is the success of the cultural shift, then it follows that great leadership is key to getting the most from the move to a digital workforce.

The *Harvard Business Review* article "Changing company culture requires a movement, not a mandate" points to the fact that we need a different style of leadership for success in a business that is transforming.[12] While the article's authors suggest that leaders can learn from the initiators of social movements who mobilize the masses to institutionalize new societal norms, we would point to the idea of *inclusive* leadership.

Jim Hemerling is an organizational change expert and BCG fellow with a focus on transforming organizations to deliver and sustain breakthrough performance. He urges the adoption of an inclusive leadership style that puts people first in order to turn adaptation and reorganization into an empowering and energizing exercise for all.[13] Central to this idea, as we see it, is that everyone within an organization should understand why they're there. This is how your initial "why" gets realized at every level. A blend of directive and inclusive leadership works best, and aligns to theories on how we best ensure that people are engaged: through leaders who have a clear vision and a road map with milestones, who then hold themselves and others accountable for the results.

How and what people manage is affected by the implementation of a digital workforce, so it's worth consciously considering the ramifications of this on the style of each layer of management. From board level, executives, and team leaders, right through to the program and project managers, the changes will be different based on where each person sits within the organization.

We've already talked about the change toward focusing on outcomes instead of outputs. Another big shift is often a move away from a "do it yourself" model toward outsourcing to a different cost center within the business. This decreases the instances of "miracle workers" within a business—those people who are the only ones who can perform a particular process or navigate a specific technology. The IT manager, too, may need to shift their mindset toward that of a vendor manager, requiring a different set of soft skills.

People management remains a core component of most organizations that have a successful digital workforce. Look again at the airline example we used earlier: once managers were used to staff who simply showed up and followed a process, but now they are managing people who have to solve problems, which requires a different—more facilitative or coaching-based—management style. Or consider the example of a data-processing team where the manager now has fewer people and a host of robots. It's a totally different dynamic. Not only that, but the leadership style required for the problem-solving (not process-following) workers who remain would also be completely different— based more around exhibiting trust, and balancing this with holding people to account.

Hard elements

Strategy is the organization's drivers: what the business is working to achieve, and how it plans to realize its mission. Strategy is about the deployment of resources; about how you're going to deliver the best outcomes toward the company's strategic goals. And because it resides over everything, it is undoubtedly going to be affected by decisions around implementing a digital workforce.

Consider governance—the separation of duties, or accountability versus responsibility. While the board may own the wider business strategy and be accountable for the outcomes of the business, the CEO and executive team are responsible for delivering the strategy within the boundaries set by the board. Governance can be at all levels, and generally allows for an overview (as opposed to being down in the weeds). It's the executive team's role to advise the board and provide them with the information needed to make decisions and set the direction. In turn, the board needs to know whether the executives are making decisions that are too tactical, and may add operational risk or become problematic to other organizational areas (as can be the case with a digital workforce). We can see, then, that a broader perspective that accounts for the wider organizational impact of automation is important.

Where your digital workforce strategy—and the associated program or project governance—sits in the management hierarchy depends on the level of adoption required. If you are just introducing a process, it may be that only the finance manager is involved. If it's part of the strategy of the organization, however, someone higher up the chain needs to champion it. From a strategic perspective, then, the board members need to increase their understanding of the implications and opportunities, as well as the threats or risks, that will come with the change toward a digital workforce. Boards often seek out risk, audit, finance, or quality-control expertise; technological understanding is often left behind. This can make it harder for board members to understand and take ownership of the strategic direction of technology. Just like with finance or risk, there will likely be a need for more technology advice or expertise within the structure of the board to provide insights and guidance around the impacts.

Structure refers to the organizational chart of a business. It's the way that the organization is structured to deliver the products and services it is known for. Far from thinking that a digital workforce slots neatly within an organization's existing structure, many implementations will fundamentally change the shape of the organization.

A digital workforce could remove management layers or hierarchies, or change the way that things are reported on throughout an organization. And who manages the robots? Who owns the team structures to deliver the outcomes from the digital workforce? In our opinion, this shouldn't be solely the domain of the IT team. It's important to see IT as an *enabler* here, as opposed to the *owner* of your digital workforce. Beyond that, though, there's the question of whether centralized is better than distributed control and ownership. On this latter point there seems to be no clear winner—the best structure differs depending on the organization, which is why it's critical to get clear guidance and a robust plan in place that suits your organization's needs.

Author, thought leader and Professor of Leadership at Harvard Business School John Kotter developed the idea of a "dual operating system" for businesses to allow them to quickly adapt to change.[14] His theory is that in our fast-paced, modern economy, businesses can no longer rely on just one organizational design. Instead, where an organization may have a traditional hierarchy or a pyramid shape, Kotter advocates for a second, more agile structure to be created to enable change. This second structure capitalizes on the entrepreneurial capability of a network and tackles rapid-fire strategic decisions for fast pivots and business reactions. In this way, the second operating structure functions more like a startup and is able to respond more quickly to change and opportunities.

In the case of implementing a digital workforce, your organization can build a new structure around the automation program. How your second structure looks can differ depending on what you've already got in place in your organization. However, this model lends itself to more of a decentralized approach. You might set up a specific team to implement RPA, or you may have each member of the automation team based in different divisions across the organization. We'll look further at the structural organization of your digital workforce program in the latter part of the book.

Whether or not an organization chooses to fully adopt Kotter's dual operating system, a digital workforce often cuts across or changes the nature of how a business is structured. In our experience, a digital workforce is typically seen as a technology project, so is often centralized with subject matter experts. Our preference is that a digital workforce is owned by the wider business, given that's where the impacts will reach. We'll include more about how to do this in later sections.

Lastly, **systems** are the organization's operational processes. We do not mean the technical systems you might be thinking of, but rather the processes and procedures that people perform—how the work is currently undertaken. It's not a stretch to see the effect a digital workforce would have on this, where what a person does is being replaced by RPA. Whether onboarding customers, processing applications, or facilitating money transfer is the aspect being automated, systems are usually the intended and primary impact of the change to a digital workforce.

So how much of what people currently do will be automated? And how much will people's activities or work processes be affected? When you implement a process, there are a host of things that relate to it that also need to have their impact considered. Perhaps our example airline might choose to have people process their VIP loyalty members or first-class flyers' check-in alongside the automated process for the masses. The exception-handlers who roam around the kiosks, which wasn't a role in the previous process, will now also need to be put in place as a result of the automation.

Automation will also have an impact on how the organization thinks about things like continuous process improvement; how you mature and evolve a digital workforce requires a different approach. For example, refining the exceptions could be part of this in the case of a digital workforce. But either way, processes—whether people-based or technology-based—often have to be improved on, or have aspects added to them, as a result of changes to systems or operations. Even in transformational leaps, small steps and tweaks will need to be made along the way.

Preparing for resistance as you manage the change

Pioneer and global thought leader on systems thinking and organizational learning Peter Senge is quoted as saying "People don't resist change, they resist being changed."[15] In our experience, this is very much true. However, we have also observed an underlying fear specifically of "robotic" automation. For many there is the apparent threat of not only being changed and all the concerns that drive resistance through that, but also of being replaced.

When we introduce digital automation, we need to remember that we are not simply changing the way someone fulfills a role; we're potentially automating a part (or all) of what they do and handing this over to a digital colleague.

Consider what happens when restructuring an organization. Often, disestablished roles have some element of work that still needs to be undertaken, so these elements will be reorganized and redistributed to the remaining people within the organization. You would hopefully not undertake this restructure without support and assistance for both the employees affected and those remaining. Significant consultation is the norm. A similar mindset is required for the introduction of a digital workforce. We need to retain a greater focus on the human element of the new organizational workforce.

When anticipating resistance to your digital workforce, it is useful to understand some of the most common causes for resistance to change so that you can plan ahead to address them. Rosabeth Moss Kanter published in *Harvard Business Review* a definitive and often-referenced list of the 10 reasons for resistance to change.[16] With a digital workforce in mind, we have revised the order of these reasons and summarize them here.

- **Sometimes the threat is real:** Taking away part—or all—of a person's role can hurt not only the person immediately affected, but also their fellow workers and friends. Resistance to the change can spread if it's not handled with due care and consideration for the individual's impact within the organization. Kanter recommends honesty and transparency to avoid what she refers to as "death by a thousand cuts." In order to avoid the latter, try to ensure that all impacts are thought through and undertaken once, rather than over multiple changes.

- **Excess uncertainty:** Kanter describes the phenomenon where people often prefer to remain mired in misery than to head toward an unknown. When introducing robots to an organization, the fear of the unknown is usually heightened. A clear, inspiring, and aspirational vision can help with this, as can having clear processes and simple, clearly communicated steps toward the vision.
- **Concerns about competence:** You're potentially taking the more mundane aspects of a person's workload from them, and leaving them with just the more challenging and exceptional aspects of their job. Even when you outline the greater creativity or scope you will be bringing to their new role, a common concern is "Can I do that?" People are, naturally, unsure about what they will have to learn and whether they will be able to do the new things that will be expected of them. Kanter recommends an over-investment in structural reassurance—providing abundant information, training, mentors, and support systems.
- **More work:** Change often involves more work—and this might be the perception during the transformation and beyond. To avoid resistance, provide the ability for the affected resources to be part of the project without overloading on business as usual (BAU) work. Often the costs associated with this are not accepted as part of an implementation budget, but it is essential if you are going to get the best use of,

and engagement from, your business units related to the implementation.

- **Loss of control of territory:** Autonomy is a motivator for people, and often implementing an automated process can remove some of this autonomy—or indeed sense of purpose—for the people affected. If unaddressed, this might lead to a feeling of "being changed" which we know is fundamental to resistance. Look to provide the ability for the affected resources to be included in planning, and provide a level of ownership of the program, where possible.
- **Loss of face:** "Why are you automating this? Didn't I do the job effectively before?" These are familiar questions when organizations look to automate part of a person's role. Leaders need to think about how to help people let go of the past—acknowledging and accepting it, while explaining the need for the future.
- **Surprise, surprise:** It is easier to say "no" than it is to say "yes," especially when you are unprepared or have something sprung on you. Changes that come as a surprise to people will often lead to many of the other resistance aspects covered here. Avoid catching people off-guard and you'll also avoid the knee-jerk resistance to change. Continuously communicate openly and seek input into the changes as you go, rather than doing all the work behind the scenes and springing it on people at the end.
- **Ripple effect:** As we've seen, change in one area of an organization can have unintended consequences

in other areas, and for seemingly unrelated people. Organizations are, at their basis, a bunch of interrelated people and systems. Ripples of disruption can cause issues in other departments . . . murmurs of "the robots are coming, we just don't know exactly where or when, but HR are using them so they could be coming here next—yikes!" Undertaking a wider stakeholder evaluation, utilizing tools such as the McKinsey 7S model and other stakeholder analysis, should be considered, with the aim—always!—being to minimize disruption.

- **Past resentments:** The specter of previous unsatisfactory technology implementations could come back to haunt the current transformation. If this is the case, consider dealing with it up front. Communicate clearly about why it will be different this time and what is in it for employees. Only elements that solve group problems will likely survive.
- **Everything seems different:** We are all creatures of habit, and generally prefer routine. Look at the level of change you are trying to achieve, and the way you are trying to achieve it. Consider the plan and look, wherever you can, for ways to minimize the impact of the changes to make the whole thing more manageable.

Thinking about the potential levels of resistance in conjunction with a thorough impact analysis can give you a better picture of the potential areas of concern for your organization. These considerations can feed into your planning for the change and help front-foot resistance as you drive toward the outcomes required.

When we introduced the *shared values* portion of the 7S model (Figure 5), we also introduced UK insurer and Blue Prism case study Ageas' experience with a digital workforce. Ageas has a three-person automation team devoted to selling the benefits and managing the message around digital change. They created videos and workshops, and facilitated ad hoc discussions to ensure that their people understood the value a digital workforce could deliver, and to eliminate the associated fears. They humanized the digital workers by giving them their own page on the company intranet, and implemented a dedicated after-care team who used dashboards and data to manage and communicate the digital workers' performance for each automated process via monthly reports to all stakeholders. And it worked.

A word about adoption

Let's pretend for a moment that we are transforming our business *without* the need for new technology. We might be restructuring or refocusing departments to work in a different way toward a new vision for the organization. It's reasonable to assume that we would want to ensure that all our employees were focused on that goal and motivated to deliver within their sphere of influence.

When implementing a digital workforce to transform a business, we need to ensure the same things. We want the human workers to adopt their digital colleagues as part of the new way of working, and be motivated to come up with better ways of doing things while utilizing their digital colleagues to achieve the vision.

It is important to understand the desired level of adoption within an organization, as it will drive how you plan and focus on change management. If you want your human colleagues to experience the mindset shift inherent in being truly invested in a new way of working, then you need to ensure that they are fully on board and engaged. They need to have a level of motivation, and a sense of purpose, to provide recommendations around further changes to existing or new processes. That's how you achieve transformation.

Herbert Kelman, the Emeritus Richard Clarke Cabot Professor of Social Ethics at Harvard University, proposed three levels of adoption that we can apply to the implementation of automated processes.[17] These are:

1. **Compliance:** We want people to do what they are told. We might assert that their job now is to process exceptions, and then use rewards and penalties to manage their workload and achievements.
2. **Identification:** We want people to understand why they are doing what they are doing. Here they need to understand their role in managing the more complex exceptions from the process, as opposed to managing the whole process as they used to. As you might imagine, for them to get to this stage, more information would need to be shared and more would need to be done to manage the changes.

3. **Internalization:** We want people to understand why they are doing this and provide input into how things can be done better. Here people will take some level of ownership of the outcome.

Stage 3—with its inherent culture and mindset shift—is what we aspire to for true transformation.

SUMMARY

- Like any business transformation, wide-ranging areas within your organization will be affected by the implementation of a digital workforce. But that's the point! If you are pursuing lasting change, then you need to have buy-in from every level of your business—and that means communication of the goals, intentions, and benefits of a digital workforce as part of your workforce operating model, especially if you want to overcome employee anxieties about their changing roles.
- At the heart, successful implementation is about the need for robust change management, especially when an automation program is intended to be transformational. Without proactively managing the change, you'll only ever achieve refinement or reinvention.

SECTION 3

Keystone principles: Increasing your likelihood of success with a digital workforce

In this section we're going to explore some of the key issues to understand at an organizational level when introducing digital technologies.

This won't be a detailed "how-to" for establishing and running a digital workforce in your organization. Instead, we'll be considering the ways in which the implementation of a digital workforce differs from software development, where automation resides in your workforce operating model, and how onboarding a digital worker compares with a human worker. Lastly, we'll turn to a simple force-field analysis to understand the resisting forces that can hinder you and the driving forces that support success in building a digital workforce.

We're not talking about software

We've made this point briefly before, but emphasize it again: *a digital workforce is not a software project, and it shouldn't be treated like one*. To the casual observer, a digital workforce can look like a piece of technology because it is, of course, driven by powerful new technology. We often create and implement digital workforces for some of the same reasons that we might put in a piece of software—to optimize a process and free up more of people's time and energy. The confusion is understandable.

The technology that underpins automation is new, which also accounts for some of the perceived overlap. Robotic process automation (RPA) was only introduced to the market around the early 2000s, so it's still in its infancy. Because of this, robust approaches and the skills required to deliver and manage a digital workforce have taken some time to evolve, although the rate at which this is happening is becoming much faster now. Unfortunately, though, because the understanding of how to best implement a digital workforce within an organization has been limited, and limited processes or methodology for adoption have existed, many businesses approach a digital workforce the same way they treat software development or other technology adoptions.

A digital workforce is a distinctly different approach from traditional software development.

> The establishment of a digital workforce is more like creating and maintaining an ecosystem or ongoing delivery program than it is about delivering a fixed-term project.

We'll explore more of the drawbacks of approaching automation like software, and the differences between the two, throughout this section.

An alternative perspective: Digital colleagues as part of your workforce operating model

In Section 1 we introduced the idea of a workforce operating model. While we're working to not treat a digital workforce like software, we should acknowledge again that it's also not a project you do in just one part of your organization.

A digital workforce is a key element of your workforce operating model. It's part of your workforce strategy, similar to the decision to outsource part of your business, or to centralize a particular business function. Essentially, these choices change the makeup of your labor force—the model of how your workforce operates. Adding a digital workforce does, too; it's not a technology project that's taking place separately or in parallel to your human workforce.

One of the authors of this book was recruited many years ago to take over a troubled technology project to deploy a new staff time-tracking solution across more than 50 sites spanning two countries. While positioned as a technology project, one point the business seemed to overlook was that as part of the introduction of the new time-tracking technology, the payroll function from every site was going to be transitioned back to the head office in each country. Besides meaning the potential loss of a key resource at each site, this also affected the comfort levels of people working in the business in regard to whether their pay would be reliably processed each week, and whether they could get answers to their questions within a reasonable timeframe. We're unclear how this could ever be seen as strictly a technology project. Do we naturally assume that just because technology is involved, a project instantly becomes a technology project despite any other ramifications?

As you'll recall from Figure 2 (the igloo diagram), the transformation that unlocks the maximum benefits from your digital workforce originates from your culture and organizational mindset. The organization as a whole needs to adopt the digital workers and accept them as the new normal, not as just a fad. It's not just about management buy-in, either—the people working *alongside* the robots need to embrace them. You want them to like their digital colleagues and come to see their potential. They will then identify new processes as candidates for automation that will have the maximum impact on their roles. Or better yet, they'll start to see whole new business activities and opportunities that weren't possible before.

McKinsey Global Institute estimates that around 50 percent of current workforce activities could be automated with already proven technologies, but that only 10 percent of these jobs are automatable at a rate above 90 percent.[18] This shows that again, we're here to *augment* rather than *replace* the workforce through automation. The need for employees is unlikely to be disappearing anytime soon.

The way forward with a unified workforce

What you are aiming for with the introduction of a digital workforce is what could be termed a "unified workforce." A unified workforce removes the friction of day-to-day activities, augmenting your workforce in the sense that technology functions alongside humans. Here, robots are incorporated into your overall workforce, referring complex cases to their human workmates to manage. They have a combined strategy to drive them forward, and are working together to support your company mission and provide better working conditions for your human team.

In most cases, a well-functioning digital workforce is not doing 100 percent of your processing; it's doing a percentage of the total volume. This works on a process-by-process basis. Some digital workforces may manage 100 percent of processes (or close to it) in the case of a really basic process that utilizes pre-cleansed and validated data. However, even a highly mature digital workforce may manage something closer to 80 percent of processing—with the exceptions and anomalies being passed to humans. And that's okay.

We need to think about RPA's capability in very different terms from that of software. You would rightly be concerned if your finance software only processed 80 percent of payroll to your staff (and so would your staff!). RPA should be designed to ensure data integrity, even while it may only process the majority of transactions and pass to humans the complex or difficult exceptions. Targeting 80 percent straight-through processing (i.e. successful results) represents a significant cost and time reduction, and a positive result in many cases.

Depending on the complexity of the exceptions, the effort required to get a robot to do the additional 20 percent of processing is often a lot more work and cost for little realizable benefit. This follows the Pareto principle, more commonly understood as the 80/20 rule, and reminds us that it's important to know when enough is enough. Many organizations make the decision not to spend additional program budget on trying to attain the final percentage of functionality when that budget could be better used elsewhere, say in deploying the next digital worker.

Let's revisit our example airline again. When the automated self-check kiosks can't cope with a differing or more challenging scenario, they will refer the customer to one of the assistants (humans) waiting patiently or working on other tasks nearby. Similarly, self-serve checkouts at the supermarket refer more complex functions to a person. Supermarkets *could* invest in the development of facial recognition software for their self-service checkouts to conduct the ID checks required to sell alcohol, for example, but right now this would be an expensive investment for minimal gain.

Do you recall the differentiation we made in Section 2 between outputs and outcomes? A technical person, who is more used to software delivery, would likely approach a digital workforce with the aim of having the technology deliver 100 percent of the transactions. Business-minded people—or those ultimately being affected by the implementation of the digital workforce—are likely to be more focused on the outcomes working, not just a specific output.

Deciding how far to take things is a tough call, but an important one to make from an economic perspective. You have to know where to stop optimizing an automated process. It's common to want to do more modifications, but this ultimately erodes the program's overall return on investment (ROI) (as in phase C of Figure 6, where more refinements are not bringing much more value).

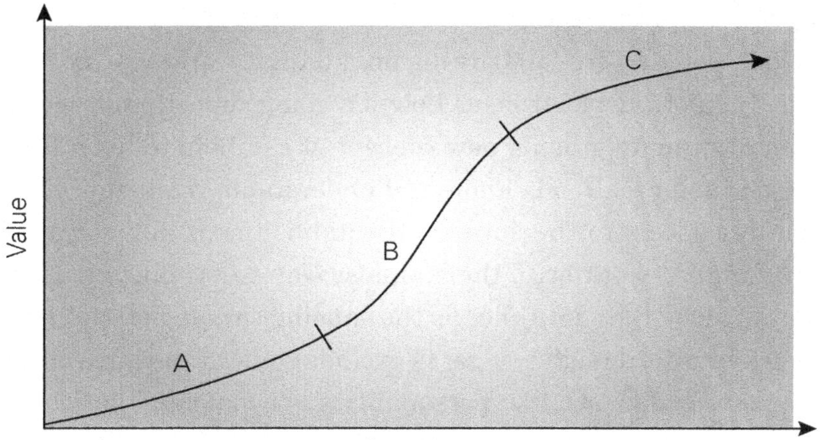

Figure 6: Cost *vs.* benefit over time.

In phase A, you're understanding your processes better, which brings a small amount of value. In phase B, you're starting to build processes and achieving more tangible value in getting results from your automated processes. By phase C, you're refining and need to determine where/if additional time will return enough value to justify it.

The cost/benefit analysis of automating additional aspects of your processes is something that can be revisited over time. As technologies improve, it is often possible to continue evolving your digital workforce—getting more and more of those business processes that are managed by the digital workforce. Expect your workforce and its cost-benefit to evolve over time—it took at least 50 years to get automated car production to the point of requiring very low levels of human-worker input.

Comparing the human and digital workforces

When you are first embarking on a digital workforce, no one really knows what it is, or how it's going to look. And when we're trying to grasp a new concept, it can help to liken it to something we already know and understand. While there are many aspects to the successful establishment and running of a digital workforce, the simplest way to visualize it is to think along the same lines as the ongoing running of a human workforce (although you get to exclude some of the inherently human challenges, like personalities and politics, that come as part of that!).

Think for a moment about your human workforce, and you'll see that there are some clearly defined—and well-understood—lifecycle functions:

- recruiting
- onboarding and training
- ongoing management.

Now imagine for a moment that you have a new role in your organization that you need to fill—let's say an Assistant Health & Safety Officer. You would start by defining the role you are trying to recruit for, and through doing so would justify bringing on board a new person—confirming the need for this role. Once the new role was approved, you would advertise the job and screen candidates. This is all part of the recruitment process.

Once you have found the perfect person and they've signed their contract, you move on to onboarding and training.

You would show them around, set some KPIs (key performance indicators) and key accountabilities, and run through relevant operational protocols. You'd have a structured induction. You would set up their log-ins and company email, provide access to (and training on) your systems; you might issue them with a device. And from there, their manager would train them up in their specific job role.

This last step is vital, and expected. The new employee may have been the ideal candidate, but this doesn't mean that they can necessarily do all of the tasks required from the get-go. The manager would give the new hire an easy job to do to start with, and check whether they can do this autonomously before adding other aspects and leaving them to get on with delivering. You'd get them there, one step at a time.

This brings us to the final stage, which is managing the employee. Here you'd be allocating work, monitoring their progress and performance, and mentoring and upskilling them as needed.

As you can see, this process is business-led; it's a joint effort that pulls together capability from HR with IT support. That said, it's likely to be primarily driven by the person or people in the part of the business most affected. For example, while a health and safety role might be driven by the HR department (as that's where the role resides), a finance role recruitment would be driven by the head of the finance department—in other words, the subject experts when it comes to finance. They're the ones who know what to look for in finance resourcing, and are the best to provide training and assess proficiency.

Now, let's contrast this with a digital workforce where there are also three defined stages or lifecycle functions:

- pipeline
- delivery
- run.

We can understand these better when we liken them to filling a human job role (Figure 7).

The pipeline stage is about identifying processes that are suitable for automation across the business, assessing candidates, and then working through justification and feasibility to gain approval. Prioritization is also key here, as—similarly to a human recruitment process—internal teams would work to determine what skills were mandatory or essential for their new worker, and which would be a bonus.

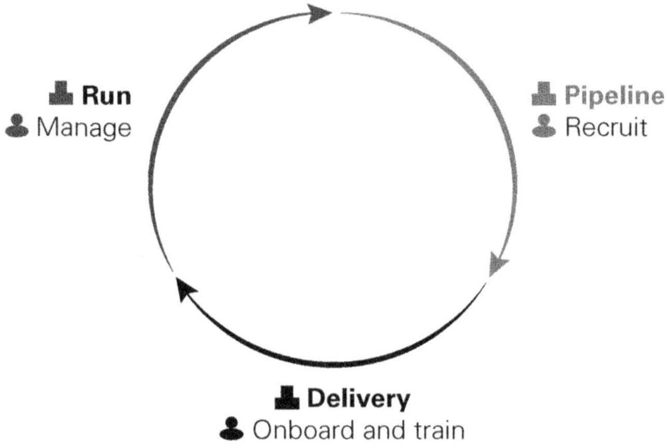

Figure 7: Analogous lifecycle functions for the human and digital workforces

Delivery, the next stage, is akin to onboarding and training in a human recruitment lifecycle. You're setting up the infrastructure and IT network accounts, designing the solution, and configuring and training the digital colleague. You're running tests as you move the robot toward becoming an autonomous worker. And lastly, you're handing the worker over to carry on its activities as part of the team.

The final stage is run, which mirrors the management of a human. As with a human worker, this is about allocating work to the robot, monitoring its performance, and resolving any issues that crop up along the way.

In the same way that the onboarding of a human worker is a joint effort across the business, this should also be the case for a digital colleague. When you train a person, you explain the purpose of the process and offer a step-by-step guide that includes all the parts of doing their job. Your trainee tries out the process while the trainer oversees and ensures that they are doing it correctly for standard transactions. In techie lingo, that's the *happy path* (i.e. transactions that don't require any exception management). The trainee then gets training, in stages, for first the standard transactions, then some of the more common exception cases—to a point—and thus gains more independence. The exact same goes for a robot.

As we can see, this kind of training relies heavily on subject experts and experienced operators carefully coaching the new person or providing input into the process that the robot should follow. They would ensure that happy paths are fully understood, validate that the robot (or person) behaves as it (they) should, and identify and trap exceptions. RPA may seem like a technology implementation, at first glance, but we're hoping it's more obvious now that this is not a job strictly for IT. Their involvement, much like for a human user, will be focused on setting up the new user, providing security protocols, and ensuring reliable access to the technical infrastructure for the user to carry out the tasks required (e.g. access to the network and applications).

Driving forces vs. resisting forces

We all know the theory of a self-fulfilling prophecy, or the virtuous/vicious circle. We expect something (positive or negative) to happen, so we act in a particular way, and our behavior causes our expectation to be realized. It's a reinforcing loop—an action produces a result, which influences more of the same action, and the cycle spirals and strengthens in whichever direction it is going.

The same concept applies to your digital workforce. There is a range of activities that will reinforce or strengthen the success of your program and move you towards successful adoption. Conversely, there are other activities that work to move the spiral in the other direction—they hinder the success of your program or take you away from successful adoption.

The idea of a force-field analysis takes this concept further.[19] Originally a development in the social sciences, and also backed by a popular change-management model, the force-field analysis provides a framework for looking at the factors (or forces) that influence a situation. We've utilized this model to apply to a digital workforce by including the specific forces that we've seen either help or hinder a successful digital

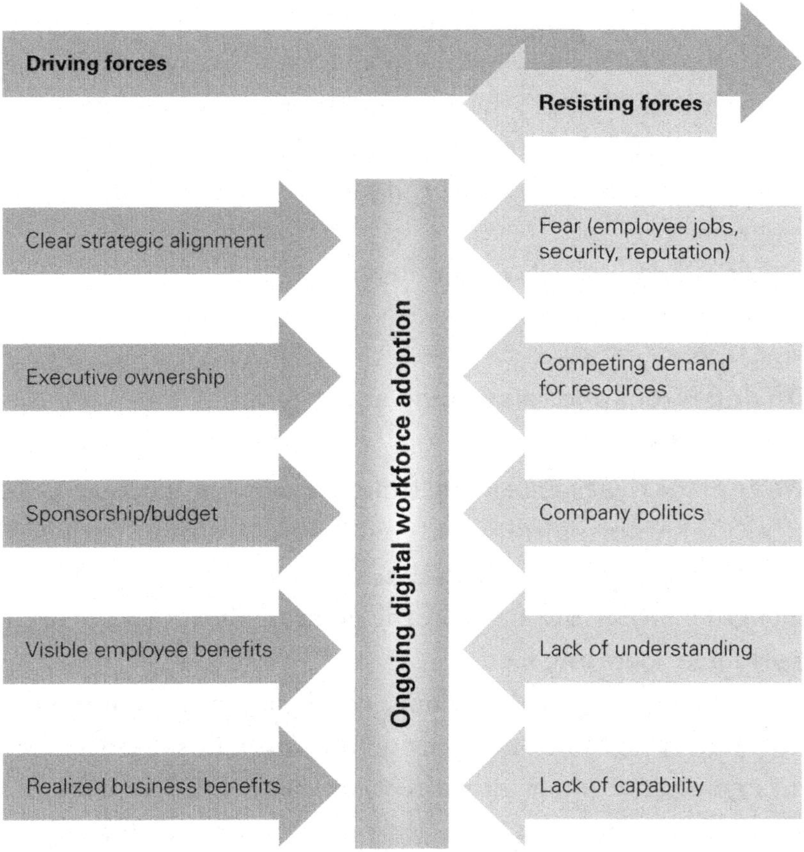

Figure 8: A force-field analysis for adoption of a digital workforce

workforce program. Take a look at Figure 8. From the left, you see the forces that are driving movement toward the goal and encouraging greater success. And from the right, you see opposing forces that block movement toward the goal.

The priority with which you need to address each of these forces may vary, depending on your organization and the specific vision you have for your overall workforce. However, we have seen that businesses that consider reinforcing activities—from overall linkage to the strategy through to communication activities and the securing of ongoing funding—are the ones that see greater success overall with their digital workforce programs.

Let's look at each of the driving factors first, then the resisting factors.

Driving forces and reinforcing activities

Clear strategic alignment

We've made this point several times already, but it's absolutely vital that your digital workforce implementation is linked to your organization's overall strategy, and that significant thought has been given to how the program fits into the overall workforce operating model.

Strategic alignment is about giving purpose to your digital workforce program. Once your organization has decided that it is going in a particular direction, clear linkages between that and the tactics to move the business forward toward its strategic goals means that those activities are far more likely to get support from all levels of the organization. When projects are aligned to the strategy, politics tend to be pushed out of the way; alignment helps to remove barriers and obstacles. Strategic alignment also helps to ensure that automation efforts get the right level of backing (with executives talking about and promoting the program) and investment (with funding assigned to support the establishment and maturing of internal capability).

Executive ownership

We've mentioned this before too, but executive-level sponsorship is vital to success—obviously in terms of getting budgets approved, but also when it comes to both the credibility of the program and overcoming hurdles within the business. We've rarely come across a technology implementation—and certainly not a digital workforce—that doesn't have some level of business implication; so success requires business leadership, buy-in, and promotion among the senior management level of the organization.

Not only are you changing the way your workforce operates by creating a unified workforce of both people and robots, but you're also trying to transform the culture and the thinking of your people across the organization. As we mentioned earlier, leaders have a disproportionate amount of influence and a significant impact on the collective mindset and culture of an organization—and, in this case, on whether transformation is effective. As their buy-in increases, leaders will model new behaviors and communicate to others more authentically and persuasively. Their advocacy for automation is vital.

Sponsorship/budget

It's one thing—albeit an essential start—to have a budget assigned to kicking off your digital workforce program, but to drive the success of your automation program the commitment of an *ongoing* budget cannot be over-emphasized. Like so much of digital workforce establishment, the budgeting process starts with setting expectations; the initial deployment of budget will mainly be geared toward building capability, as well as starting to get some deliverables.

Unrealistic expectations or KPIs mean that from the outset you're setting the program up to miss the mark—especially as there's a period of ongoing maturing required. If the wrong executive team—say one without full backing for or understanding of the program—got hold of it, moves could be made to cut funding or move it to the next "shiny" thing. One team told us that a lack of understanding about the need to invest in capability up front led their organization to get rather lost in the hype. Within two months of this team being established and beginning to move up in maturity, the executive team was expecting to start seeing AI integrated with their automation—which the implementation team was never going to be able to deliver within such a short timeframe.

Because capability isn't readily available within most organizations, success requires a commitment to some level of consistency. Once you've taken the time to build up your internal capability, or secured the best people from a vendor's team, you can't turn things off and back on again and expect the same standards and efficiencies that you were starting to develop initially within the program.

Visible employee benefits

We've seen already, with our emphasis on the importance of change management, the need to get your existing human workforce on board with automation. Employees will be expected to work alongside their digital colleagues, to handle exceptions, etc. As we will explore further in our final section, we want employees to see the automated processes (robots) as *colleagues*. Ideally, people from these teams have recommended the processes to be automated and are helping to come up with new ways to use the technology. If they are not seeing the benefits, they're not going to support it—which ultimately leads to the pipeline drying up.

We have shown you many of the benefits of automation already, but many of these apply at an organizational level. So, what benefits can your people expect to see at an *individual* level?

A successful digital workforce frees people up to work on more of the high-impact aspects of their jobs. It can enable them to do things more autonomously, and to take on more challenging roles that require upskilling—which is inherently motivating and bodes well for driving the success of the digital workforce overall.

We think this point is worth a little history lesson. In the 1970s, Edward Deci, a psychology graduate student at Carnegie Mellon University, set out to expand on earlier studies on motivation. He found that human beings have an "inherent tendency to seek out novelty and challenges, to extend and exercise their capacities, to explore, and to learn."[20] This invalidated longstanding ideas about what drives us as people. The previous—and in many ways still pervasive—belief was that beyond our biological need for survival, the best way to improve performance, increase productivity, and encourage excellence was to reward the good and punish the bad.

However, consider open-source software, or the development of Wikipedia that succeeded where MSN Encarta did not.

People pour incredible amounts of their time, energy, and expertise into developing, writing, then offering up their outputs for free. Sure, in some cases these people might bolster their reputations or sharpen their skills, thereby enhancing their earning potential. However, it's not simply, or always, a case of financial reward. There is also intrinsic motivation at play—people do it because they are driven by a sense of purpose and because they enjoy it.

More recently, MIT management professor Karim Lakhani and Boston Consulting Group consultant Bob Wolf surveyed over 680 open-source developers and found that "enjoyment-based intrinsic motivation, namely how creative a person feels when working on the project, is the strongest and most pervasive driver."[21] In freeing up people from mundane and repetitive processing tasks, and allowing them to be engaged in more creative and complex activities, a digital workforce encourages this.

American author and motivational speaker Daniel Pink was across this with his theories around autonomy, mastery, and purpose, and now more and more businesses are realizing, too, that this is where a true competitive advantage lies—in better inspiring and enabling their people.[22] This is a huge employee benefit.

A digital workforce can allow people to take more control of their careers and work-lives while also helping organizations retain a talented workforce. A government department we once did some work with is an excellent case in point. This business unit had a team of very smart people carrying out vital analysis. Their researchers would collect numerous data points, then their analysts would release final figures once a month. A more-detailed release appeared each quarter. The team hated the final part of the process and the system they had to interact with. They loved their jobs—aside from this day each month, which was incredibly stressful as there were very public deadlines that had to be met. Automating the final part of this process wasn't so much about making significant time savings; instead it was about not wasting these clever people's time (or stressing them out)—and it ensured improved staff retention while establishing increased certainty that the public deadlines would be met.

Realized business benefits

It is pointless having a robust, adaptable digital workforce that has achieved a positive ROI, is meeting service level agreements (SLAs), and is scaling to meet variable workloads, if you're not communicating this within your organization. And it's never too early to start. One major banking client of ours set up automation dashboards from day 1—even before their first process was built. This meant that as soon as the first digital colleague went live, they were already showing updates and communicating to the business. Since then, employee and customer benefits have been identified, tracked, and communicated via the company-wide intranet and in executive reporting. Demonstrating the organizational value that their digital workforce has delivered has made a significant difference to the business's success, and has enabled them to cement support from their executive team. Such transparency remains key, especially if you are working to transform.

Transformation requires an organization-wide mindset and culture shift, one that inspires people to think about the opportunities of new technology and look for ways to constantly improve or innovate. After all, transforming a business involves transforming individuals; an internal shift has to occur.

One of the important activities for promoting greater automation is seeing the automation that you have in place performing well. Whether it is the cost of the automation delivery and its running costs far outweighing the previous human operational costs (from wages through to recruitment, training, facilities, and support), or the ability to manage volume variability or seasonal spikes being realized, these achievements can all be used to make a case for more automation. Other benefits, such as increased adherence to compliance requirements or a greater consistency of service achieved, also help to strengthen the ticks in the box for automation.

As well as a positive ROI, achieving—and making visible—ongoing efficiency gains will also be a significant driving force for automation. There is an overhead involved with the configuration and set-up of a digital workforce, so it's important that this is reduced and that delivery is optimized as the teams conducting the build become more experienced.

Further efficiencies can be found in the reusability of developed components that can be utilized for future processes, and the use of purchased pre-built components or processes to support more-efficient delivery and scaling. We've seen countless examples where a particular process might be cost-prohibitive if you were to build all the components from scratch, but can be built viably using components from a similar process that has already been created. Demonstrating this to the business builds momentum to keep your pipeline of candidate processes healthy, which is a key determinant of success.

Digital workers create efficiency, and these efficiencies can enhance your customers' experiences. Sometimes you can discover this with internal metrics. You might know that the throughput of a transaction once took ten working days to process from start to finish and now your digital workforce is able to do it in three, so customers are able to get their claim paid (or whatever the result of the process is) faster. In other cases, a customer survey might provide solid evidence of customers who are happy with their level of certainty or with response times. Either way, this reinforces a positive case for expanding your digital workforce. After all, what company doesn't have an organizational goal around improving customer experience? Just ensure that you make these improvements visible throughout the organization, for maximum benefit.

Finally, your digital workforce, once established, might support other key strategic objectives of the organization, such as increasing customer satisfaction and providing a consistent and fast turnaround of customer queries or requests, or providing the ability to scale services based on variable and unpredictable demand. Consider the case where an organization has outsourced a function of the business offshore; the resource may be affordable, but it's inconsistent and difficult to manage. RPA can support the transition of this function back, to instead be carried out by a digital worker.

Visibly delivering to other strategic and tactical objectives such as this within the business can be a significant driving factor for your digital workforce.

Resisting forces: Obstacles to success

Where the driving forces strengthen results, buy-in and success, resisting forces do the opposite. These are the obstacles or inhibitors of success, or the reasons that we most often see digital workforce programs fail.

Before we get into the specifics of what the resisting forces are, we want to point to the potential here of a lag. Inhibitors can come with a time delay. You can launch your program with a hiss and a roar, and have it appear to be going well, but if you're not doing the right things (i.e. proactively focusing on the driving forces to counteract the resisting ones) at some point your pipeline will dry up, and the program will lose support, engagement, momentum, and, finally, investment.

Fear

As well as needing to get clear on and communicate why it is you're implementing a digital workforce, there's an education element that's often required—the priority of which is to remove as many fears as possible about the changes. This can only be achieved with transparency and authenticity. We've heard of several instances where organizations have started a pilot around a digital workforce and then just continued on, with little or no organizational communication. And all the while, they're slowly laying people off. Or, alarmingly, we've seen one case where people were coming in to meetings about the automation project in tears, believing that the program was going to take their jobs away.

A successful digital workforce cannot be run by stealth. This will only breed fear and rumors, and severely undermine trust. In this scenario, there will be few, if any, suggestions for new processes to automate ventured by the business; and when you're evaluating processes or working to configure new ones, you're likely to encounter and receive only substandard information from the people affected.

Instead, be honest about why you're implementing a change, reassure people about what's not changing, and work proactively with your people to find purpose. Share the successes of the program with the rest of the business to inspire them to start shifting their thinking. Transparent communication can play an important role in bridging the gap between a new technology and the rest of the business.

Speaking of technology, challenging the security of the digital workforce can be a significant roadblock—and tends to also require education. Understanding how the robot or robots function can decrease fear and paranoia significantly. If this is new ground for a business, new security considerations need to be considered, including segregating the duties of those configuring the digital workforce from those who are supporting and running the digital workforce. In a leap that's hopefully not too difficult to make by now, we would advocate for the security of the digital workforce to be based on the same sort of security framework you would put in place for humans performing that same function or process. At this point, it's also worth noting that configured digital colleagues are less likely than humans to have a bad day and break the rules, or do things they should not be doing.

Finally, the digital workforce needs to establish and maintain trust with the business. We've seen that executive-level support is vital to success, but we've also seen examples where the senior people in an organization steer clear of automation in order to protect themselves and their reputation because they don't have 100 percent confidence that it will work. It's critical to find people who will believe in the program and support it wholeheartedly, and while some of this needs to happen up front, the program must also continue to prove itself by way of steadfast delivery.

When it comes to maintaining trust, there are a couple of elements. The first relates to data integrity. This seems like an obvious point to make, but it's important that all transactions processed by the digital workforce are 100 percent correct. Always. This is not the same as 100 percent of transactions being processed—it's that the ones that *are* processed are always done correctly. This requirement is sometimes why people decide that "enough is enough" in terms of refinements—it is pointless to jeopardize the integrity of your organizational data by attempting to squeeze just a small percentage or two more of performance from your digital workforce.

A common perception is that you can actually trust a robot more than you can trust a human—after all, human errors are removed. But your digital workforce has to show, first, that it works how it should and that it will keep performing—which brings us to the other element underlying trust: robustness.

Reliability and consistency of your digital workforce's

processing is vital. Just like any business unit, your digital workforce needs to be a dependable function within your business. Robustness relies on adherence to SLAs, which are akin to KPIs from a people perspective. In the same way that it's important to not put in place KPIs that drive the wrong types of behavior, it's vital that our SLAs (say, around speed of processing) don't undermine the accuracy or reliability of the digital workforce.

Robustness and reliability breed confidence—for the executive team, and for anyone within the business. It's all well and good to have things working when your robot first goes live, but that robot needs to show up and keep performing—not stop working the next week when something goes awry with the server. To risk stating the obvious, unreliability will severely inhibit the success of your digital workforce program.

Competing demand for resources

Internal competition for resources—budget, people, and capacity for the project—is a vital consideration as it can lead to resisting forces if not managed. For starters, although digital workforce programs should not be primarily driven by IT, the involvement of IT people is still required. Those IT people, however, have other projects on the go, and often conflicting priorities. We've seen sticking points where IT constraints hinder the setting-up of the program, so that delivery teams are not able to get on with building robots and further establishing the digital workforce.

Another aspect is making sure that you are working with multiple parts of the business, and that those departments are supported in terms of their involvement in the program. Concentrating your efforts in just one part of the business often means that the team is setting the pace—the automation program can only go as fast as the affected team can onboard it. Although the introduction of automated processes shouldn't be overly burdensome on the team adopting them, they will need to spend some time with the delivery team.

In the same way that the finance function might not recruit a new employee right in the thick of end-of-financial-year activity, while running application upgrades, or during another period of greater demand as they won't have time to onboard them, the same goes for implementing automation—the requisite time and space are needed. A department might want the automation, but ideally the internal people affected need the support of the organization as a whole so that they have the room for learning that can decrease resistance to change.

Company politics

Internal disputes over ownership and controls, or a lack of clear governance, are detrimental to the success of a digital workforce program. The first way to counteract this resisting force is, as mentioned earlier, to ensure that your program has strategic alignment—a clear link to the wider strategy is likely to trump office politics each and every time.

Another factor at play here is that often automation is about relieving some of the most pressured areas in your business—that's one of its benefits. However, this depends on the people involved, and their wider view. We've seen situations where one business unit is monopolizing more than its fair share of automation efforts and trying to optimize processes that aren't, overall, all that problematic. While this might help those individuals to meet their personal targets, it's not the way to maximize the program's ROI.

Clear ownership shows who is responsible, who is making the decisions, and whose KPIs are being affected; we'll elaborate on this when we explore the ownership pillar of our digital colleague adoption framework in the next section. However, whether relating to aspects of ownership, organization, or control, it's vital to have the goals of the digital workforce take precedence over individuals' own interests.

Finally, IT and the business need to be on the same page when it comes to a digital workforce and why it is being adopted. There needs to be alignment. Many of us have experienced the seemingly age-old divide that can exist here, but this needs to be bridged for a digital workforce to ultimately succeed. Some conflict can seem inevitable, but both parties need to appreciate that this is new to both of them and to work together; if there are existing conflicts between these groups, these need to be resolved first. We have seen delays in the provision of infrastructure or system accounts for proposed digital colleagues that have had major impacts on the efficiency of the delivery of the digital workforce program, as well as a significant impact on the ROI of the process being configured.

Lack of understanding

Understandably, a general lack of knowledge about what is required to establish and run a digital workforce inhibits success. It's a mistake to view the building of digital workforce capability strictly as a software delivery project, when it requires much more big-picture thinking. The business of automation has multiple bedfellows. Any organization looking at building a digital workforce should be considering HR, internal communications, risk and change management, as well as IT, to ensure that a digital worker can achieve what it needs to.

Treating RPA like an IT project results in not fully understanding the business processes, not developing a large pipeline, and not being able to optimize the selection of processes to automate (which enables you to take maximum advantage of reusability). Combining a business-led approach with an IT lens ensures that the organization will have considered the functional requirements (i.e. the specific process that will be automated), as well as the non-functional requirements—the criteria that ensure your digital workforce will be able to perform those functions effectively. You need to consider questions like these:

- What security protocols need to be put in place?
- How have back-ups and platform maintenance or updates been addressed?
- Have the availability levels required of the digital workforce platform been considered?
- Does the robot still run if some aspect of the platform, or an adjacent system, falls over for whatever reason?

This level of understanding is essential for success, and IT can help the automation champions within a business cover more of these bases.

> **Ideally, automation teams need to work closely with the business's technical people, while not being solely composed of them.**

We've seen already, too, that missing the internal change-management element is a severe hindrance—it misses bringing the people within the business along on the journey, and means that achieving the mindset shift required for true transformation is impossible. As part of that change management, internal marketing is important; as is education.

A misconception that we encounter frequently is the expectation that an automated process will perform much faster than it does. However, an automated process still has to emulate the same steps that would be undertaken manually. And in most cases, there are two parts to the way a process is performed: the human action and the system's action. While the emulation of the human action performed by the digital colleague can be made to go faster, the response times of your existing applications are not going to be sped up.

Consider searching for something on Google. You (a person) type in the search term you are interested in and ask the site to do the next part. The system then takes over, going away to retrieve the required information before refreshing the page to deliver those results. As we've said, automation generally replaces (and may increase the speed of) the human action, but it doesn't speed up the application that a person (or robot) interacts with. You will typically see a speed increase in only 50 percent of the steps in a process. Sometimes you won't even get that, because the automation needs to logically check more aspects than a person might need to. It pays not to get too hung up on speed, however—when you consider that a robot takes no coffee breaks or sick days and can work 24/7 (instead of nine to five, five days a week), there are still plenty of benefits overall.

Lack of capability

It's incredible how many digital workforce programs are being put together without the requisite skills being present. And the results are poor. Not having the right people on the bus—i.e. with the right skills—will affect most areas of your digital workforce program.

To have any hope of successful adoption or transformation, you need to have the skills and capability to educate your existing workforce, promote automation within the organization, and run your digital workforce. Otherwise, it's likely that not only will the wrong processes be selected, but you also might see poorly defined and configured automation, and instability. Or worse: incorrect results. RPA isn't just anyone's game. Too many organizations are throwing juniors with a manual at the high-stakes task of building digital workforce capability. Many think that a short online training course equips them with all they need to know to get this right. When people without the right skills launch into RPA, you unfortunately end up with inefficient and ineffective automated processes that are badly designed, cannot be scaled, and don't deliver the intended business value. And it's easy to see that this doesn't set you up for success with your automation program.

If a digital workforce's infrastructure platforms have not been designed, established, or supported in a robust and maintainable manner, this will have a negative impact on the ease with which you can scale them up as part of expanding your digital workforce. Poor design and configuration will adversely affect robustness, performance (in terms of speed and volume of successful transactions), and the reusability of the digital workforce's components. Ultimately, this erodes the efficiencies that are possible. Your infrastructure is your backbone; so from day 1, you need to spend the time required to build a strong infrastructure and processes that are capable of scaling and being agile.

Substandard infrastructure can cause technical issues and potential outages as platforms are maintained. Meanwhile, a digital workforce that is poorly designed and configured increases the level of maintenance required as each automated process is delivered. Think ahead of where you are and what you're building now. Even if you don't yet have the maturity for certain aspects of scaling, it pays to have your eye on the future from the outset. At the very least, identify what you might need and have a plan for what might happen as you mature.

While no single part of building digital workforce capability is especially difficult or complex, there are so many small pieces that are absolutely vital to get right. And because the thinking around RPA is still in its teens, there's a real shortage of people with enough experience. The right advice and focused expertise are still vital. Organizations would also benefit from adopting more of the website or .net practice of downloading ready-made components or processes that exist for RPA, as opposed to building everything up from scratch. This is an evolving space, with the likes of RPA-Marketplace, the Blue Prism Digital Exchange, and UiPath's Marketplace taking the first tentative steps into this exciting space.

Business-led IT security and support

As we discussed earlier, the establishment of a digital workforce is part of your organization's workforce operating model. It's like onboarding, training, and managing people to conduct an activity within an organization. Therefore, we hope it seems logical that since the business best understands its processes—and the upstream and downstream dependencies and other implications—it is the business that should own the establishment of the digital workforce.

When we refer to the business here, we mean largely the users—those people that will be affected by the digital workforce. These people will have the best understanding of:

- the SLAs
- the details of the issues that might be encountered with each process
- the spikes of demand for each process, and
- the future direction and life expectancy of each process.

As internal communications and marketing are essential for the successful adoption and scaling of a digital workforce and the reduction of fear, these functions should also be squarely the domain of the business. We have witnessed how the business is far better equipped to do this than the IT department on its own.

That said, there is an important role for IT to play here. Just like they would support the induction and onboarding—and the ongoing work-life—of a human workforce, IT needs to look after the digital workers. In this instance, IT would set the security standards just like they would for a human—although these are often less rigid than what is required for human workers. The same goes for infrastructure: IT are responsible for the establishment and ongoing support of the digital workforce platforms, in the same way that they would look after the infrastructure of your enterprise resource planning (ERP) system or what your main operations or finance systems run on. Your digital workforce will likely be running mission-critical processes, so it's vital to have the appropriate levels of support, high availability, and disaster recovery controls implemented.

However, now we're getting slightly more into the details—so we'll pause here and dive deeper in the next section, as we start to explore our final model: the Digital Colleague Adoption Framework (DCAF).

SUMMARY

- A digital workforce is not a software or IT project, but instead an extension of your workforce operating model. Ideally the aim is the creation of a unified workforce and in fact, humans and their digital colleagues have more in common than many originally think.
- Like with a typical self-fulfilling prophecy or reinforcing loop, a digital workforce has a range of factors that strengthen or support its success and another subset of activities or factors that can hinder or work against successful adoption. Linkage to the strategy, proactive communications and ongoing funding and executive support may all drive greater digital workforce adoption success, but they also must counter inhibiting factors like fear, company politics, and a lack of understanding.

SECTION 4

The Digital Colleague Adoption Framework (DCAF)

Up to this point, we've looked at the opportunities and implications of a digital workforce in terms of its potential for transformation. Following the keystone principles, and asking the essential questions of yourself and your organization, will have set up much of the high-level thinking that is key to successfully implementing change at a strategic level. Now it's time to roll up your sleeves and start getting into more of the detail. It is worth noting that this framework is not intended to be focused at the delivery level; instead, its aim is to provide a wider organizational view of how best to incorporate your digital colleagues into your organization's workforce.

This section covers the practical activities that need to be considered and undertaken to set you up with the best possible chance of success in introducing a digital workforce to your organization. In Section 3 we looked at the various resisting forces that prevent an organization from deriving the maximum value from their digital workforce. We also started to explore the positive driving forces that counter those obstacles. Here we will take many of those ideas further.

This section is not a step-by-step guide to implementing a digital workforce. There are as many unique implementations as there are organizations, and as such the practical activities will vary. What we offer here is a framework built from our extensive experience of seeing what works and where the challenges lie, as well as our experience of being hands-on in helping organizations overcome the hurdles to success. This framework—and this section—will support your organization to consider and implement the right processes to best realize the value of automation.

A word about frameworks

In essence, a framework is a structure for support. It's the basis of a project. It's a set of assumptions, concepts, values, and practices that constitutes a way of viewing reality. It concerns itself with how parts fit together.

It is not, however, a guarantee of success. After all, no framework in the world can provide *that*. It is the *implementation* of a framework that provides or drives its success (or otherwise). We've seen a lot of frameworks fall short. Some people would even assert that frameworks don't work. That may seem like an odd thing to say, especially as we're about to offer you a framework. But the point we want to make is that they *can* work; they simply can't do it alone.

By necessity, a framework must be understood within a context: the context of the opportunity, the context of the technology, and the context of your strategic direction. And this is why we are only getting to the offer of a framework here, in Section 4. First we had to lay the groundwork of all of that thinking you've been doing in the earlier sections. Without that, a framework wouldn't be of much use. You need to be able to look at how it fits into a business. A sense of purpose is what makes a framework successful; you need strategic linkage.

In Section 2 we looked at a project-management approach as compared with a more change-management-based approach. Where the former focuses almost exclusively on outputs and measurable advantages, the latter looks at those outputs in action and tracks the adoption of the technology to deliver the benefits. Change management also looks at the wider aspects that a project might overlook if it is strictly output-driven, like culture and people and the—perhaps unexpected—influences of implementing something new and different. And that's the clue to why we see many frameworks lacking in effectiveness. Taking a more holistic view from the outset, one that incorporates change management principles and softer skills, is essential.

A brief history lesson

When RPA first began gaining traction in businesses, the major providers developed excellent models to assist organizations to implement this new technology. Misconceptions were—and still are—rife. People were muddling through with something new, and without positive use cases or as-yet robust ideas around "best practice" they were learning as they went. So, these frameworks were hugely valuable in helping organizations get the building blocks in place and implement a new technology in their efforts to achieve greater efficiencies.

However, as was necessary at the time, these first frameworks were largely focused on implementation, with lots of detail about delivery. This was a time when few people knew how to go about implementing RPA successfully—what documents to use, how to validate the solutions, or how to set up the technology. Very few yet had the requisite skillsets. These frameworks looked at *how* to deliver the technology. It was about getting the technology set-up right, and less about the long-term building of benefits.

The Digital Colleague Adoption Framework (DCAF)

Now we are moving beyond the "how." We've seen more businesses bring in this new technology, and there are more people with experience at this level. More recently we've seen the emergence of the driving and resisting forces we explored in Section 3, and there's increasing evidence of the limitations and scaling difficulties of *not* incorporating change management.

The 2019 McKinsey Global Survey found that although more organizations were starting to see success with digital workers, the biggest challenges to adopting automation were still managing employee resistance to change, and attracting talent (meaning the capability for delivery). "The fear of the future can be a powerful deterrent," the report asserts.[23] This in no way downplays the importance of extraordinarily strong ability within the delivery team—that still is the second of the challenges, after all. However, McKinsey predicts that by 2030 up to 375 million workers globally will need to change jobs or upgrade their skills significantly. Businesses therefore need to recognize their responsibility in taking the lead and helping their people make this transition.

In order to take their people on the journey and have

them truly adopt a digital workforce into their organizations, businesses need an approach with an onus on managing the changes within the business that are brought by the new technology. The DCAF builds on and extends the thinking of earlier models. It's a framework for the business, not just the program or technical team, to ensure that organizational considerations have been taken into account to provide the best opportunities for success. We've strengthened or extended some of the aspects of those earlier models—the primary difference being the inclusion of the human element as new non-human workers are introduced into the fold. Thus, the DCAF is not just about how you implement robots within your organization; it's also about how you educate and develop your current (human) workforce to embrace the benefits of digital workers, work alongside them, and promote unified workforce opportunities.

The change-management element is crucial. Your existing workforce needs to be considered, informed, educated, and managed, in order to truly maximize the benefits of the digital workforce and to increase adoption, efficiencies, and business value. The value that this extended model provides is what makes this section of the book so important.

WHAT'S IN A NAME?

Before we move on, we want to share how we named this framework. Automation is intended to work with the rest of the business as a colleague, co-worker, buddy, teammate, ally. Indeed, the idea of a "digital colleague" is that it's something you work *alongside*, rather than add-on software or a plug-in that checks your grammar. Then there's the purposefully chosen term *adoption*. When you adopt a dog, it becomes part of your family. In the same way, when you adopt your digital workers they're more likely to be accepted and endorsed as part of your organization.

And finally, our acronym (DCAF) for the Digital Colleague Adoption Framework was also deliberate. Or we could call it a "happy accident." Decaf coffee helps you get stuff done without stressing you out; it avoids heart palpitations and doesn't keep you awake at night.

A look at adoption and maturity, and how change management fits in

We know that introducing a digital workforce is not a case of simply flipping a switch. Think about onboarding a new colleague, or the establishment of any new operating function within your organization. As an example, let's consider the creation of a new, centralized procurement function. As part of this journey, your organization would have to go through many maturity and adoption phases as both the business unit itself matures, and as the rest of the organization learns and adapts to working with this new department. This new function would likely take some time to establish, and then even more time to fully implement the processes, controls, and communications required to meet the specific needs of the business. The establishment of a new digital workforce program within your organization is similar, although potentially even more difficult.

The distinction here is not intended to understate the complexity of procurement; rather to suggest that as a function it is considered a matured professional practice. Procurement has a legacy of people understanding its purpose: the managing of purchasing for all other divisions, and negotiating new supplier agreements on behalf of the business. RPA doesn't yet have that legacy. There is a shortage of seasoned RPA professionals, and while your organization is likely to have people who have worked alongside procurement teams in the past, it is unlikely that many people in your organization have worked alongside RPA teams. Less common, too, are people with experience working alongside digital colleagues. These situations represent changes that need to be managed.

In observing the many RPA programs that we have been involved with, we can say with certainty that the process of maturity takes time. After all, RPA is still in its teens! From the initial set-up of an automation team there is the steep learning curve of delivering the first few processes and taking these from requirements, through configuration and testing, to deployment and production. It's a process of continuous maturation—usually over a couple of years—as the delivery team and the business team come to understand RPA and what working alongside automated processes really means. They come to see what change management is required through the organization, and the adaptations required to existing processes and controls from an IT and risk point of view. It's a learning process, and it doesn't happen overnight.

An organization has to work through these phases of

maturity, which mirror those related to adoption that we saw at the end of Section 2.

- **Initialization:** This first stage includes setting a vision for RPA within your organization, establishing the capability, and upskilling the required resources to deliver digital colleagues. You're just getting started. You're dipping your toe in and you won't have yet seen the business value; it's still mainly about potential. From an adoption point of view, this is the first, or *compliance*, phase. As far as your people go, this runs along the lines of "I'm doing this because we are trying it out."
- **Industrialization:** Next there is a scaling out of digital workforce delivery to multiple departments within the organization. Here you're working to secure organization-wide buy-in and a strong pipeline of candidate processes with the potential for automation. This stage usually involves increasing the team dedicated to your digital workforce, and bedding in best practice and more efficient delivery. More people throughout the business will be actively adopting their digital colleagues. We're moving now into engagement and, from an adoption point of view, *identification*. For your people, this tends to look more like "I'm doing this because I can see the value."
- **Institutionalization:** And finally, this is where the organization reaches a state of thinking "digital workforce first." The use of robots for conducting

processes becomes as common as the practice of using humans. In some organizations, we've seen this happen as a freeze on recruitment until it has been determined whether a digital worker could fulfill (or partially perform) that role. Another organization has the automation program report up to the vice president of HR to help facilitate this approach. When you consider that hiring is one of the more expensive functions an organization undertakes, the choice to look to robots first can present a better value proposition.

In our procurement example, institutionalization would be when the procurement team successfully become top of mind and the default behavior becomes "Oh, I'll go and talk to the procurement team about this first." This is full *internalization*; it's full adoption. Here we have people in the business engaging in a way that shows commitment to the new way of doing things.

Maturity and adoption truly are key to realizing the full potential of a digital workforce. At one end of the continuum, the robots are just there, they exist; at the other end, your people think of them as co-workers and consider how they could work better together. By its very nature, adoption infers acceptance. Welcome to the family. It's about taking something on board, and it represents an element of choice. It's much less about being forced to do something; it's an option. And that sort of embracing of change—in this case in the form of digital colleagues—is really what we're going for here.

In "Hyperautomation: The most significant RPA trend of 2020," George Lawton urges organizations to "think deeply about employee engagement in automation initiatives."[24] In this same article, P.R. Krishnan, Executive Vice President at Tata Consultancy Services Ltd (a global leader in IT services, consulting, and business solutions) is quoted as saying "We have seen repeated success with automation efforts when employee engagement is fostered and there is a participatory culture."[25]

The point here is that you will never get to the final stage of maturity, nor a scaled, successful digital workforce, if you don't bring your people along on the journey with you. As you'll recall from Section 1, it is not possible for transformation to take place unless your people are right there with you. As the adage goes: If you want to go fast, go alone; if you want to go far, go together.

Not all organizations need to get to the institutionalization stage to benefit from a digital workforce. If you're not adopting automation with the intention of transformation, then reaching industrialization—or even, in some cases, initialization—is enough. However, if you do want to transform, you most definitely need to institutionalize. That's what enables you to develop this potential powerhouse to its full value. We love the idea we've seen in a couple of organizations leading the way—the concept of an "always on" transformation, which points to the continual nature of increased adoption and maturity. It's a mindset geared toward continued change and evolution ahead of a fixed outcome or future state. It offers more organizational agility and makes a lot of sense; especially when you consider the time that the journey to adoption and maturity takes, the continually changing environment that an organization exists within (recall our igloo?), and the cyclical nature of the automation lifecycle.

Performing a pilot

We've seen many organizations pilot the use of RPA technology within a segment of their business before committing to a full digital workforce program. This is sometimes referred to as the "inception" phase, preceding the three aforementioned phases of initialization, industrialization, and institutionalization.

Because the technology has already been proven around the globe, some vendors claim that pilots are an unnecessary waste of time and money. In our experience, we still believe pilots to be worthwhile—unless you consider them in the strictly traditional sense. Here's the rub: If you're doing a pilot as a functional test—just to see if the technology works as it should and integrates with other applications—then you're probably investing unnecessarily. The value of a pilot for a digital workforce lies in looking at the organizational impact, and especially how it's going to affect your workplace culture.

Because all organizations are different, it's important for each to specifically understand what their structure and capabilities mean from an adoption or resistance perspective. A pilot allows organizations to proactively plan for how they will manage the changes a digital workforce will bring. It also offers people within the business a "lower stakes" opportunity to start getting comfortable with the technology and what it will mean for them. It's testing the waters before plunging in. From a change-management perspective, a pilot can be used to help you identify the innovators and early adopters in the business, who may be good targets for the start of a full digital workforce program.

The idea of piloting a program to test it out and shore up uncertainty has its merits, even from a communication theory perspective. The late sociologist Dr. Everett Rogers furthered the more contemporary version of the "diffusion of innovation" theory as it relates to change and the way an idea gets promoted and adopted through social systems. Rogers suggested—and quite rightly so, we feel—that people have to understand and trial a new technology in a safe manner before they are ready to move to fully adopting it.[26] A pilot provides another opportunity for doing this.

It's also an opportunity to assess vendor fit. Most organizations embarking on automation utilize the skills of expert vendors for either ongoing (or at least initial) support. The pilot phase then can be as much about validating the vendor's capability and trialing the working relationship, as it can be about understanding the implications for the business. And this makes sense if you're looking for a long-term partnership.

A pilot should be about gathering information; it allows your organization to understand how it can maximize the benefits of a digital workforce, and how it will impact its workforce and customers. It can also allow you to determine or refine your vision for a digital workforce within the organization. You'll see more about the merits of this shortly, but first let's take a closer look at the framework itself.

DCAF—the mechanics of the model

Our Digital Colleague Adoption Framework (DCAF) provides a basis for the important decisions you need to make:

- *What* the organization is aiming to achieve
- *Why*, as part of your overall organizational workforce operating model
- *Who* is going to sponsor, govern, promote, deliver, and monitor the digital workforce
- *Where* the automation will be conducted (from a physical, operational, and structural perspective), and
- *How* this will be delivered, as well as how the aims will be achieved from a technology and existing workforce perspective.

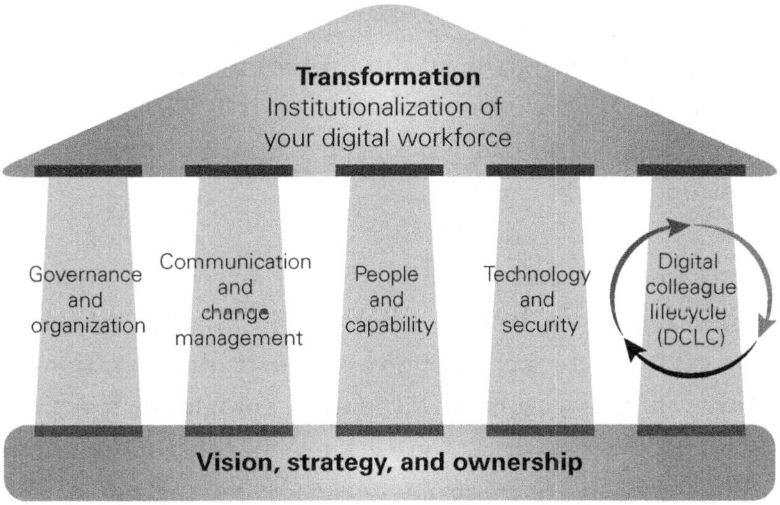

Figure 9: The foundation and the five pillars of the Digital Colleague Adoption Framework

The framework consists of five pillars sitting on the solid foundation of ownership. They are:

1. **Governance and organization**—governance of the entire digital workforce programme through planning, delivery, and ongoing running of your digital workforce.
2. **Communication and change management**, and how this important aspect will operate on an ongoing basis.
3. **People and capability** required to implement a robust digital workforce within your organization.
4. **Technology and security**, and the involvement of IT in governance, infrastructure ownership, and security.
5. **Digital colleague lifecycle (DCLC)**, which concerns itself with all the stages in screening, creating, and running a digital workforce.

As more and more artificial intelligence (AI) becomes incorporated into our business operations, a sixth pillar will likely need to be added in—one that considers legal and ethical issues concerning how AI is used.

In the same way that you can't have a roof without the walls, it's the strength and robustness of the five pillars that holds up the overall success and adoption—or institutionalization—of your new digital workforce. If you want automation to achieve success and scale, and to effectively drive transformation within the business, attention to each and every one of these pillars is vital.

From our experience on-site with many businesses

embarking on their digital workforce journey, we have seen various frameworks utilized. We've also seen mixed levels of adoption and success. Although how best to implement the specifics of a framework will vary from business to business, these pillars remain important across the board. Where organizations have omitted one of the pillars, the result is slow adoption; or, in some cases, a total halt to their digital workforce program—and not because of a lack of suitable candidate processes to be automated. Instead, holds have been due to aspects of the framework not having been thought through or suitably implemented. This then impacts on delivery efficiencies, the transparency of benefits, and organizational-wide buy-in, as well as executive-level support.

The five pillars can be worked on progressively and in parallel. However, like good walls, you can't have any significant gaps. The place to start—unsurprisingly—is the foundation. You can't have walls that stand straight and sturdy if they're not built on a solid foundation, so you need to think about vision, strategy and ownership first. The way that you then think about the pillars—and the level of thinking and decisions required to optimize the speed and smoothness of your adoption of a digital workforce—all depend on your vision. A vision of transformation while aiming for institutionalization will have a very different methodology to how you might approach your pillars for a simple technology implementation. In other words, you need all the walls, every time, but the foundation determines how tall they need to be.

So, let's look at the elements of the framework in deeper detail.

The foundation: Vision, strategy, and ownership

By this point in the book, you'll be aware that you need to think about *why* you want to implement a digital workforce—and have a vision for it—up front. We've spent plenty of time on this aspect throughout the earlier sections; you could say it's been a foundation of this book!

If you want a business initiative to succeed, then you need a clear **vision** that aligns to the overall organizational vision and is incorporated into the organization's strategic plan. In the case of a digital workforce, this needs to be incorporated into the strategic plan as part of the organization's workforce operating model. You might be considering a digital workforce as a growth enabler, for increased agility, to retrench, or with the intention of utilizing robots alongside natural attrition in the business. You may even be thinking of it as a foundation to future aspirations of AI.

It may appear that we're taking you back to Management 101, but this is because it's vitally important that the vision and goals for your digital workforce are clearly articulated and communicated. As well as alignment to your organization's overarching vision, the vision for your digital workforce needs to incorporate transformation plans (if you're going to that level). Either way, you need to consider not only how the digital workforce will be utilized, but also how your people will be prepared for change, educated, or given other opportunities. It seems obvious, but we have seen too many organizations that aren't doing this well.

As we briefly touched on before, we've witnessed one organization which—while in post-pilot mode—was still trying to implement RPA by stealth. It didn't take long to reveal how badly this was working. The business analysts were confronted by a team of people they were supposed to be working with, suddenly all in tears. They were afraid that the analysts were there to give their jobs to robots. This wasn't a good situation for anyone from the business or the delivery team. It didn't bode well for internal marketing and adoption, either. From a practical point of view, not having and communicating a clear vision is a hindrance to getting clear and accurate requirements from the users, which can lead to expensive delivery and multiple changes required to the digital workforce.

If you have a good idea of your vision for your digital workforce, you can use your pilot to augment and develop this further. If you're not 100 percent confirmed regarding your vision and ownership, a pilot can be utilized to help you refine or change aspects of this so that you can move forward from solid ground.

Similarly, we've covered aspects of **ownership** in the earlier sections—and the message remains the same here: clarity is vital. Due to its transformational nature, a digital workforce needs executive-level sponsorship and ownership. It needs someone (or a team) high up who is responsible for making the program work.

We've mentioned before that the vision for digital transformation needs to be created and owned by the business, not by IT. Outsourcing the ownership and accountability for the digital workforce away from the business, away from the people who make it possible, just doesn't work. According to the McKinsey article "Four success factors for workforce automation":

> One of the worst pitfalls we have seen during the implementation of automation at scale is when executives view automation as another IT service, to be led by the IT department. Even if automation is assigned absolute priority, day-to-day hardware and software business requirements will swiftly push automation to the back of IT's queue. Responsibility for automation must ultimately lie with the unit being transformed, not outsourced to an often-overburdened IT function.[27]

And we agree. The finer details of how a digital workforce program is structured, and where the delivery of it sits is where we head next as we look at our first pillar: Governance and organization.

First pillar: Governance and organization

This pillar is about where the overall digital workforce program sits within your business; how it will be structured, delivered, and managed; how it interacts with other divisions within a company; and how it will be funded or charged to business units.

Where should we start automating?

One of the early considerations is in which area, or areas, of the business you are best to introduce digital workers. In other words, you need to define the initial internal target customer of the automation. Ideally, you want keen business leaders and teams who will be supportive of the initiative—not just from a business unit management perspective, but also in terms of the culture and adaptiveness of that team.

It's helpful to remember the typical innovation adoption curve here (Figure 10),[28] and that different people embrace change at different times or to different levels. Early adopters within your organization will want to be some of the first in the program. Conversely, some of your more conservative players and departments are unlikely to want to be the front runners; they'd rather be left for later when they can see the proof of successful automation working.

From a practical perspective, understanding this allows those implementing a change to target the creation of "critical mass" for that change, which improves adoption. Targeting the innovators and early adopters, followed by the early majority, helps do that, meanwhile engaging in additional communication to reduce resistance from the late majority and the laggards. It's important to think not just of the

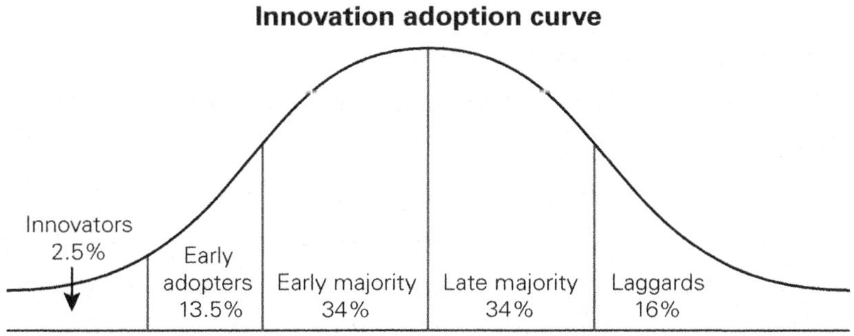

Figure 10: Typical innovation adoption curve

business unit managers and where they fit on the curve, but also where the people in their teams sit in relation to adoption.

We've worked with one innovative client who went to HR to determine the areas of the business with the highest staff turnover. He was working on the theory that there were likely to be mundane or unappealing aspects of these roles that made retention difficult, and these could therefore be targeted as the first potentials for automation. Wherever you start, it pays to "share the love" a bit across business units that meet your criteria: it allows your program to maintain momentum, reduce risk, and avoid missing other opportunities. As we've seen, too much focus on one area of the business can overload their capacity and means that the whole program is stalled if there are restrictions to that business unit—whether from unexpected extra demands on their time, a change of direction, or even something like a seasonal rush.

That said, *too many* areas can also be ineffective. Spread yourself too thin, and it's tricky to get traction. Efficiencies come from working with teams that come to understand the delivery approach, as well as from the reuse of components that's more likely possible when you're involving similar or overlapping business units that access the same business systems.

Following the selection of the initial business unit (or units) for automation, a plan is required to determine at what stage (based on timing, benefits triggered, or number of processes that are live) the scope of the digital workforce will extend to other business units. Ideally, the order in which adoption is extended would take into account the readiness and willingness of those other business units—their innovativeness and the ease with which they adopt technology. This maps out the process of how you will get to critical mass; all the while focusing on other reinforcing activities, too, such as celebrating successes and the achievements of the early adopting business units. We'll talk more about this in the discussion on the communication and change management pillar.

How does the digital workforce team fit into your structure?

When it comes to structure, there are almost as many possible scenarios as there are organizations. We have seen huge variation in where the automation function sits within the business or IT, including centralized, distributed, or a blend of the two. The success (or otherwise) of your structure depends on your specific organizational culture and on whether there is a close and trusted working relationship between business units.

The decision regarding where your digital workforce program sits has a significant impact on the overall success of your program. It will influence the level of support for the program within the wider organization, the confidence people have in the delivery of it, and the trust people have in both the solutions delivered and the ongoing running of the function. Getting it right matters.

We would advise starting with a centralized model for delivery capability and then expanding. In George Lawton's aforementioned article on hyperautomation, Martin Fiore, one of Ernst & Young Global's managing partners, is quoted as saying that "Overall, having one group focused on RPA is helpful in driving collaboration, sharing experiences and broadening the community to attract talent."[29] We've seen examples of this sitting within operations, as a stand-alone business unit dedicated to business improvements, and, in one case, reporting up to the vice president of HR. As Fiore goes on to add, it pays to have a mixture of business and technology expertise in that group. Later, even if you do move to a distributed model, ensure that the quality and standards capability remains centralized—as with an automation center of excellence—to avoid compatibility challenges and inefficiencies.

While the delivery of your digital workforce undoubtedly requires expert capability (as part of an upcoming pillar), capability also has to do with how you structure and organize your digital workforce program. Unfortunately, automation is often positioned as having a low training cost. We've seen far too many cowboys jumping on the bandwagon, or organizations thinking that they can train up their internal people easily. While it is possible to turn exceptional business or IT implementers into excellent automation implementers, we haven't seen great examples of this happening. More organizations start out with external vendors who have proven experience, then train and transition the skills to internal teams.

The important thing to remember here is that it's the individual consultants' skills that are important, not just the track record of the company they represent. To help organizations get to the heart of this, one of our colleagues has developed a capability testing approach that individuals from various vendors can be put through before commencing a project. This is another instance, too, where performing a pilot can come in handy: it's a good test for validating your outsourced talent's skills, as well as how well they will be able to work in with your organization.

The benefits and drawbacks of outsourcing should be considered for your specific organization; however, one of the dangers can be unwittingly outsourcing your IP. Some organizations get around this by using their own business analysts to assess, define, and re-engineer processes, then outsource the capability to deliver them in a more robust manner. This is a good approach—it retains specialist knowledge within the organization, as the IP tends to reside more in the deep understanding of the processes themselves than in the technical configuration.

Governance issues

Regardless of where the digital workforce capability resides within the organization, a well-defined and well-functioning governance model needs to be implemented. Containing common goals for the successful adoption of your digital workforce, your governance model should include the business's end-users, who will be working alongside their digital colleagues, IT, HR, and the communications or marketing team.

For your overall digital workforce program, a governance model would encompass:

- setting goals and KPIs to ensure that the digital workforce program is established robustly, and set to meet the required organizational vision for automation
- ensuring that the correct reporting and mechanisms to measure benefits realization at each stage of the Digital Colleague Lifecycle (DCLC) are in place
- sharing in the communication of your program's successes across the wider executive team
- managing the removal of roadblocks which would hinder the effective delivery of your digital workforce
- ensuring that the digital workforce program is sufficiently and consistently funded, so it's not something that is stopped and started again
- implementing an organizational change management and communications plan correctly, to keep your wider audience informed and to mitigate fears
- ensuring the inclusion of your organizational HR, and learning and development teams to proactively retrain or cross-skill your human workers, who could be displaced by the onset of automation.

Further understanding of the governance functions as they relate to a digital workforce can come from a look at the stages in the DCLC.

Governance in the pipeline stage

While the delivery team is responsible for the collation, assessment, and recommendation of candidate processes for automation, it is the governance function that is ultimately responsible for setting the selection criteria and approving, deferring, or rejecting processes for your program on a case-by-case basis. The governance function is also responsible for ensuring transparency of the criteria and selection process. Failure to do this will only lead to reduced organizational buy-in and a diminishing pipeline.

Here, the governance team needs a solid understanding of the information being presented in this stage so that they are able to make informed decisions—factoring in process size and costs, risks, support requirements, and benefits.

Governance in the delivery stage

Although a delivery team may take the lead here, the governance function continues to play a major role. The governance team takes responsibility for ensuring that:

- the delivery team are obtaining the support of business units they are working with
- the required capability exists within the delivery teams
- delivery targets are being met, and
- early identification of risks and appropriate mitigation actions are being implemented.

Governance includes ensuring that the correct levels of project and external controls are implemented from a risk and compliance perspective. This encompasses internal and/or third-party security assessments of critical or high-risk processes, and making sure that the correct levels of user acceptance testing and sign-off are implemented.

Governance in the run stage

Governance here spans two major streams: the technology view, and the operational view. From a technology perspective, the governance role ensures the correct technology support, care and preventative actions, as well as ongoing security reviews and capacity planning so that automated processes continue to run robustly.

From a business perspective, governance is responsible for ensuring that SLAs are understood, that the underlying technology is able to support these, and that there is tracking in place to ensure these are being met. Governance is also concerned with ensuring that the correct level of operational monitoring and reporting is implemented, and that prompt resolution of processing issues is occurring—as well as whether business continuity plans have been put in place to support the processes and the human workforce alongside them. The exceptions to the way a process runs are likely to factor in at this stage also. You'll recall how we said that most digital colleagues deal with the majority of processing, and elevate a small percentage—the exceptions—to humans to handle? Governance considers the wider workforce managing the entirety of the process (of which digital colleagues may make up only 80 percent or so).

Second pillar: Communication and change management

This is one of the areas that our framework adds onto those that have preceded it. By now, it will be no secret that we believe this is one of the most important aspects of a digital workforce for organizations—and the most often overlooked. We believe that it is also one of the most common reasons why organizations struggle to scale their digital workforces.

In one of the in-house business units we used to work in, we adopted a saying: "We're not here to do things *to* the business, but *for* the business." This was needed to counteract the way that many business units perceived IT. It is also vitally important here, as a digital workforce often has to overcome the same sorts of perceptions and work to establish trust.

If you're not actively promoting what you are doing, you are swimming against the tide. It's a natural aspect of human behavior to be resistant to being changed. We also seem to be instinctively wired to complain to friends or colleagues (or whoever will listen!) about bad service we received—much more so than mentioning good service that we've encountered. Therefore, unless you do internal marketing to actively promote the successes of your digital workforce program, you risk the murmur of opposition getting louder and negativity becoming the prevailing narrative. Internal promotion that is both factual and educational is key.

Change management and communication ensures that your people will be taken along on the journey, and that it will be a positive journey for them. It's important to not leave any areas of the business out, either; this will raise the possibility of fear about "When is this going to happen to us and what will that mean?"

The most successful digital workforce programs we have seen include a continual communications function that provides the whole business with transparent updates on the state of their digital workforce. They'll even talk about the bits that may not have been going so well but they are working on improving. They provide dashboards to management and other teams to show how their digital workers are performing on a day-by-day basis, as well as tracking the incremental benefits over time (usually showing how many hours per day of productive time have been returned to the business by each process). Regular updates are usually provided in internal comms, newsletters, and via company briefings. One technique we have seen work very well is the presentation of videos of an automated process running, helping to educate the business as well as providing processing reassurance.

All of this works to establish and increase trust. And trust is not just related to the integrity of data (i.e. the digital colleague doing the things it is doing correctly and reliably). It also comes from transparency within the business and frequent, authentic, and proactive communication. A lack of openness and transparency about your program, and a lack of communication, stand in the way of success.

> At the end of the day, digital workforce success is not about technology deployment, but about managing complexity and cultural change.

We've talked about this as it relates to transformation, and in the earlier pages of this section we saw McKinsey's take on the challenges organizations face.

Automation changes the capabilities required by people to perform their jobs. It changes the nature of the work that people are currently executing.

The opportunities of a digital workforce are clear, but success comes down to the way the transitions are managed by organizations and their people. The McKinsey article "Four success factors for workforce automation" goes on to say: "We believe the most useful way to view automation is as a tool that puts innovation at the hands of employees."[30]

Businesses need to start gearing up their whole organization for adaptability. It's not a case of simply telling finance—or the business function that is being affected first—that some of their roles are changing.

> Adopting a long-term view and providing education about what's coming will promote readiness across the whole organization.

In order to truly have digital workers adopted as colleagues, the people working alongside them need to have their fears removed. Understanding precedes engagement, and it's common that we fear things we don't really understand. That's why a big part of successful adoption and scale comes down to good change leadership and communication.

We've seen some organizations do this remarkably well, such as running competitions around naming their robots.

One organization put on an "innovation showcase"—a multi-day event for everyone in the organization to look at the cool stuff that was being done internally. There, visitors could see a life-sized cut-out of Campbell, one of their robots, and were even encouraged to take a selfie with him. The organization built awareness, removed the sense of threat, and encouraged conversation and adoption.

Another aspect is the importance of people feeling as though they have been considered in the change conversation. One of the key things encountered again and again in change management is a reminder that we can't underestimate the power of people's attitudes and beliefs. Some of these can only be unearthed as you embark on solid change management. As an example, one bank adopting automation believed that they were making positive moves to remove the repetitive and "boring" aspects of tellers' jobs. However, those people on the frontlines of the bank believed themselves to be performing a valuable service. They saw those tasks as vital, as helping them to meet their purpose. Effective change-management communication needs to acknowledge that not everyone shares the same subjective experiences. There's likely to be resistance in taking away something that others perceive as meaningful.

Ideally, as you work through an effective change-management and communication program, your people will have their input sought. They will come to see how automation is aiding both the organization in achieving its vision, and they themselves in getting to do more interesting work and pursuing their own individual purposes.

Change management in the pipeline stage

If it wasn't already apparent in our exploration of this pillar, organizations need to start thinking about, and implementing, change management as soon as they start an automation program. This should be a consideration even while you are undertaking pipeline analysis. You don't need to have all the information to hand to make a start. If automation is going to happen, let people know and then keep them up to date with further communications. Starting early enables you to get early—and stronger—buy-in for the selected automations. It enables you to define why the change is needed and creates a positive sense of urgency. It enables you to include possible resistors.

We've alluded to this already in our previous pillar, but a key consideration here is how you evaluate the readiness of the business unit you're dealing with. For example, you don't want to start implementing a digital workforce into a part of the organization that is already experiencing significant change. Like you looked at the broader organizational implications in Section 2, using McKinsey's 7S model, it's good practice to now apply that framework to the affected business units and understand the impact of introducing automation.

As well as identifying business units, the other important activity is identifying the key stakeholders across the organization—particularly in the business units first included—and preparing and including them from a change-management perspective. You need team leaders and managers of the area you are working in to be well educated and on the same page regarding a digital workforce.

Start communicating the program and pipeline as early and as often as possible, as we've said. It's important to promote automation while being careful not to overpromise or hype it. We see this happening all too often. Bob De Caux, Vice President for AI and RPA at ERP provider IFS, has similarly asserted that he feels challenged having to navigate the hype around RPA when working with enterprises. Instead, there should be the expectation that automation will represent a period of organizational learning, that a program will need to bed in, and that the organization will need to mature it.[31]

Transparency is key, and the education process starts here in the pipeline. It's about setting expectations. Automation is different to software delivery, where the IT team goes away and comes back with something all done and ready to put in. Speak to the engagement and involvement you will need from the business, as well as the timeframes involved. People within the business need to know that this won't be a case of getting a new digital colleague next week, just like you wouldn't expect a new human worker to start the next Monday after you've just identified a need for their role.

A final note as we move through the lifecycle: just as your pipeline itself is continuous and evolving, your change management and communication around the pipeline should also be this way.

Change management in the delivery stage

Unlike the broader business scope of the change-management activities in the pipeline, those within the delivery phase target the specific areas of the business, and the processes being worked on at that point in time. The focus here is more on the individuals being directly affected, and their needs, while keeping an eye on the wider impacts of a transformative technology on the rest of the business.

Plan your change-management activity as an integral part of delivering digital colleagues, and start managing the change at the same time as you start the delivery phase. Further your stakeholder analysis and impact assessments of the specific automation change, and ensure that the change scope is clear—and clearly communicated. It's important that communications also include what is *not* changing, and what the expectations of the human workforce will be following implementation. Your human workforce will likely need to undertake the more difficult aspects of the processes being automated, so motivation and upskilling requirements or further training should be planned, all while being mindful of the impact on "business as usual" throughout.

The focus at this stage will be on getting business users working closely with the delivery team to allow for regular updates and demonstrations, as well as the ability to provide feedback. This all helps to alleviate fear and create more buy-in, but also points to the importance of your delivery team being good people managers. The delivery team needs to think about how they are dealing with the people in the business, and do this with empathy. It's important to understand that the people you're dealing with are going through a major change. All of this is likely to be new to them, even if you may have been involved in delivering automated processes in the past.

Finally, balance the bigger picture, even while you are dealing with the detail. Continue linking what you are doing to clear business goals, and stay focused on clear objectives. You want to do all this while also capturing metrics about how you're going about doing the delivery—with the aim of continuous learning and improvement. This is where the Agile methodology can come in handy (see the Resources section). Showcase what you have achieved in the last increment of delivery (also known as a sprint) and have a retrospective debrief with the business around what worked well, and what didn't work.

Change management in the run stage

While the establishment of the pipeline is a continuous activity, the delivery function works on processes as projects before handing them over to run teams. Because processes can experience a few teething issues when they are first run (this is true of both human-led and automated processes!), it's important not to take your hands off the wheel here when it comes to change management. Instead, look to keep up communication about a particular project until it is bedded in well to the run area. This means that the run stage consists of both project-based communication and more general promotion of the overall run function.

As well as process-specific communication, it's vitally important to publish the success of automation more generally across the wider organization. This should not just be a case of sending out correspondence to those departments you've been working with. It's also about getting the word out to those divisions you haven't yet started automating—as well as up to, and through, the management layers to ensure that the program continues to get onstream support at the executive level. Continued publication of automation successes from the run function is a key aspect of the business delivering a continued pipeline.

When it comes to reporting and communicating about performance, it's important to get your facts straight. Track the performance of your processes so you can publish the benefits realized in a way that ensures reviews are based on facts, not perception. In doing so, you'll avoid that all-too-common creep towards negative perceptions over time. One way could be to capture and socialize specific metrics, such as how many exceptions occurred and whether that number is decreasing.

Third pillar: People and capability

We've said it before, but we'll say it again:

> While no single part of building digital workforce capability is especially complex, getting all of the aspects right is the difference between simply automating processes and realizing true business transformation.

How you structure your digital workforce team will usually be influenced by the scale of the digital workforce delivery function within your business. We've seen separate managers and teams responsible for each of the three lifecycle functions (pipeline, delivery and run) in some larger programs; in smaller programs, the functions may share some team leaders.

To follow, we list some of the traits and skills that are essential to each of the three functions or phases. As you read this, it's likely you will see a very clear message: that while you do need the correct technical skills required for delivery, the key traits are people skills, people skills, and people skills.

Even though we are talking about digital workforces, we are actually talking about the people around them.

Before we get to those aspects of the lifecycle, however, we'll outline some key capabilities when it comes to the leadership that makes a digital workforce successful.

Capability in an automation manager

We know that great leadership makes or breaks a business, and here we're talking about the leadership behind scaled capacity. No one blinks when you talk about the challenges of scaling a startup, and we all inherently understand the notion that what works with 10 to 20 people is unlikely to work when you've got 50-plus. We also understand the idea that we need to mature our workforce to avoid stalling under the weight of admin, sales challenges, and disorganization as we grow. Scaling a startup requires disciplined leaders and experienced business-people.

Scaling a digital workforce is really no different.

When you're implementing a digital workforce, it can be advantageous to have management that "gets" the startup mentality—people who are rolling with the punches and celebrating the small wins. They're hungry for better. They're able to see opportunities where others might have missed them. Ideally, an automation manager isn't just working alongside a strong and strategic vision within the organization—they're also leading with intrapreneurial discipline and keeping an eye on the future.

As is the case for success anywhere, a symbiotic relationship between strong leaders and a strong team is essential. That's because the right leadership sees a digital workforce as a strategic solution, not a tactical one. When you have this mindset from those leading the program, it changes how and why you're operating. And we know that if you get buy-in and engagement at the executive level for digital workforce implementation, then you're better able to empower business transformation at each and every level.

Capability in the pipeline stage

As to be expected, it's critical to have people in the pipeline team who are good communicators, and who understand both the business users and the requirements of your digital workforce as it aligns to the vision and strategy of your organization. Key at this part of the lifecycle are people skills, the ability to educate others, and empathy. The people responsible here must be credible—they are critical for promoting a positive image of the digital workforce, which includes communicating the benefits and alleviating fear, while also being realistic about what a digital workforce can and cannot do.

From a more technical standpoint, the people in this part of the lifecycle most often possess some business analysis skills that enable them to review existing documentation and determine the suitability of candidate processes for automation. This is great because many of these people have existing relationships in your business and they understand the upstream and downstream processes.

Capability in the delivery stage

Delivery teams are usually composed of business analysts and process designers who can understand the current business processes and define the design for the required solutions. What's critical here is that at least some of this team should have a high level of skill in the configuration of digital colleagues. The more substantial automation programs we have been involved in also have a "design authority" function to ensure the robustness of designs and check that the delivery people are following standards and best practices. This is critical for ensuring that robust and maintainable solutions are delivered.

When it comes to dedicated testing capabilities within delivery, there seems to be a variety of levels of involvement. Some teams rely on the skills of their delivery experts to validate solutions with the end-users, while others have very formal testing practices.

One of the key things that we hope has come across in this book overall is that this is not a time to just send a few of your business or IT people on a training course, or have them complete a short online course, and then expect them to be able to perform this function well. While a digital workforce may appear relatively easy to get up and running, the devil really is in the detail. Automation requires practiced, specialist skill.

It takes experience to understand what you need to put in place to ensure that your delivery is robust, reliable, auditable, and maintainable. In the same way that you would have HR specialists dealing with people, you should have a digital specialist dealing with automation.

Unfortunately, we have seen many examples of people diving in after a short course—and even situations where juniors from a vendor have been used to define the set-up of an entire digital workforce program after simply reading a manual. You can guess how this turned out for the vendor and the customer!

Capability in the run stage

The people assigned to this part of the lifecycle need to have both a high level of specialist digital workforce understanding, and great communication and interpersonal skills. After all, they work with the business teams, the IT function, and the delivery team. These people are usually referred to as *controllers*, as they schedule, monitor, and reallocate resources to ensure that processing volumes are managed.

We've seen the ratio of controllers to processes vary hugely, depending on the robustness and reliability of the digital workforce, the stability of interacting applications, and the run practices and tools implemented. In one of our enterprise banking clients (where much of the administrative role of looking after the robots has been automated), one person easily manages a digital workforce that is running more than 50 different kinds of business processes spanning a number of different business units across the organization. This far exceeds the industry norm of one controller scheduling and managing five to 10 processes. The former requires a different mindset, tools, and approach to what is possible, but it also brings with it a significant cost saving and the ability to scale the automation program without the associated scaling of run resource required.

Fourth pillar: Technology and security

We're truly not trying to downplay the important role that technology and IT plays in the successful establishment and ongoing operation of a digital workforce. Technology is vital to overall success, which is why it is one of the vital pillars holding up your program. Right from the outset, technology plays an important part in setting up the whole system and making sure the infrastructure is correctly in place.

Getting this technical foundation right—the correct set-up of the underlying infrastructure to support your digital workforce—is as important as the essential network technology that allows all your people to interact with each other, your customers, and your suppliers. It has the same level of importance as core systems like your ERP and CRM that are vital to operations. Your digital workforce may well be running mission-critical processes, so it can't afford outages. It's vital that the right level of high availability and disaster-recovery processes are put in place and validated. Vital, too, is the right level of monitoring, maintenance, and support.

It's worth also giving some thought, at this stage, to how your digital workforce might scale. We mentioned the startup mentality before; it applies here, too. Your infrastructure is your backbone, so it's important to build a strong platform and processes that are capable of scaling quickly. You need this ability in any startup, and especially in the case of a digital workforce. The infrastructure should be set up to scale quickly and easily when a new digital worker comes online, or as workloads rapidly change. This gives your digital workforce the organizational agility it needs.

One of our banking clients has managed this exceptionally well. Not only are they able to monitor increased process demand across their team of digital colleagues, but they can easily switch their infrastructure to meet demand. For example, one platform becomes an investment-management running machine for the day, rather than a loan-processing one, in response to spikes in the market. Because the processes are already configured, it's simply a case of scaling them up as needed to change how the business responds to different types of demand. This is much simpler than trying to deploy human resources to do a job in a different department if you suddenly needed them to.

The way we see it, there are three key aspects to the involvement of technology in a digital workforce.

The first is **education**. It's important that the technical people in the organization understand RPA technology. Those contributing to the digital workforce—from a security or infrastructure viewpoint—should gain a solid understanding of RPA, and its differences from software and other technologies. When we have initial discussions with an organization about a digital workforce, there are commonly a lot of fears around security. Most think that a digital workforce requires more restrictive security, because they don't yet fully understand how it works. When we've explained how the technology underpinning automation works, and how the robots interact with other elements, the perceived high risk decreases and a more rational view is possible.

Importantly, this changes the way that the technical resources in the organization approach setting up their robots. We find that organizations that approach set-up with overly dictatorial security requirements struggle to get their processes working smoothly, as they have to set up different types of network accounts, which causes a range of other challenges. These challenges can be minimized with an understanding of the right levels of control to be applied to the management of the digital colleagues.

The second aspect relates to **governance**, as we explored in the first pillar. Here we will merely reaffirm that IT people need to be providing input when it comes to governance.

And finally, there are the activities performed by **technology**. These span:

- the establishment of the underlying infrastructure for automation, from servers and databases to virtual desktops
- security
- monitoring and maintenance plans
- high availability and disaster recovery (there could be mission-critical processes being automated)
- how the delivery team would request things like network and application access for the robots

- considerations around scale
- who is going to be doing what within IT, and
- how the delivery and run teams, and the business, are going to interact with IT on the above points.

Technology in the pipeline stage

The role of technology here mostly lies in evaluating the feasibility of candidates as you work through process selection. You'll want to know about the robustness of the applications that automation will be interacting with, and whether there are any concerns around interacting with those systems when it comes to things like performance, stability, and load. You don't want to kill any applications, after all!

Next, it's worth considering the lifespan of a particular application. It may be useless to build something that uses the CRM if the business has already made the decision to significantly upgrade or replace it. Timeframes factor in here. If an upgrade is targeted for 12 months away, it might still prove worthwhile to implement automation in the interim, with the intention of updating this later. It's helpful to know when updates are going to happen, or what the update cycle is for applications that your digital colleagues might be interacting with. Understandably, this has a bearing on the amount of maintenance you will need to do on that automated process going forward.

And finally, you want to know what environments are available to your applications. Say you have a process that you want to automate. Because of licensing, there may not be a test or development environment available—only a production one.

In other cases, the infrastructure to run a test environment might be expensive or cumbersome to set up, so it's important to know what versions and environments are available to use for automation. It's also important to understand what data is in those environments, including both how current it is and what sort of sensitivity the data has.

Technology in the delivery stage

The most important point we want to get across here is that technology's initial involvement for delivery is not all that dissimilar to setting up a new human user. Like a new human hire performing tasks, digital workers need access to a production network account, and the right levels of access to the applications they are going to be using. Where a new hire might require access to training environments, the delivery team will also need accounts set up in a development environment where the digital worker can be configured (or trained) and tested prior to being carefully released into the production environment.

Also, much like any new hire, sound security is still paramount. Our recommendation is to set up the security for a digital workforce in much the same way as you do for your human workforce.

Following the configuration and testing of the digital worker, it's time for it to be set to run against production systems, and alongside other users in the business. While it's easy to think that the deployment of a new digital worker into your production environment should be conducted in the same way as deploying a piece of software on your systems, this would be a mistake. It is important that IT governance and managers fully understand the differences between a digital worker and other software, as the traditional management practices currently in place to release new applications into the business won't work with digital colleagues. It would be absolutely taboo to think of making changes to a software solution in the production environment—but for robots, this may legitimately be required in many cases. In fact, it's common to need controlled and short-term access to production environments for your digital colleagues.

We can imagine some readers starting to break out in a sweat about now—but once you understand the controls around this, you'll realize it's not much different to a team manager updating standard operating procedures that their team uses to conduct some processes. In fact, it's safer and more traceable than the human version of this.

There are cases where digital workforce processes can't get configured in a development area. For example, there might only be access to a particular application in a production environment. And this is okay, although it does have to be done in a controlled way, usually with a subject matter expert sitting right beside the developer. In a recent presentation at the 2020 Blue Prism World online conference, CTO at East Suffolk NHS Foundation Trust, Darren Atkins, explained that due to extreme time pressures during the COVID-19 pandemic, they had no choice other than to build all their new processes directly in production—with obvious controls in place.

This is like training a new human user in a live setting: the trainer would sit with them, or accompany them to perform their new task for the first time, allowing for feedback adjustments along the way. Using an RPA platform where the user can easily see, understand, and validate what the automated process is doing, and whether it's conducting the desired activity correctly, also helps.

Technology in the run stage

Let's start with what we already know and understand. When digital workforce solutions have been validated by the business, made ready to go live, and moved to a production environment, some aspects of the process are similar to traditional release management. As an example, there's still the need to have a level of segregation of duties to ensure the releases are robust, and to restrict developers from having totally free access to production environments.

Most other aspects are very different from traditional software release processes, as we've touched on. In traditional software development you package up your software and deploy this into a quality assurance (QA) environment, to test both the final application, and the package and release process. Due to small platform and system differences, this may not work for digital workforce processes—it is akin to training a new user in a test environment that is different from the production system or live instance that they are going to be using as they actually do their job.

Also, there are often cases where minor RPA configuration tweaks could be required in a production environment; these would be validated by the subject matter experts following the release to production, thereby allowing for a minor difference between the QA environment and the production environment.

This should not include changes to business logic, and should be managed in a controlled manner.

With a digital workforce program, the turnaround time for changes being able to be deployed into production is also quite different from standard internal software release procedures. When the digital workforce team requests that a small change be deployed to production within a couple of hours of being tested and approved by the business users, it may be tempting to conclude that you have a bunch of cowboys running the digital workforce program. In reality, we must keep reminding ourselves that a digital workforce is not a software solution and needs to be treated differently. We have heard of major banks implementing a 30-minute turnaround of digital workforce issue resolution—i.e. identifying an issue (which could be an unexpected change to an application the digital worker interacts with), making changes and testing with the business subject matter expert, and then deploying back into production in a controlled manner all within just half an hour.

From a technology perspective, ongoing support of your digital workforce is the same as supporting a human worker. Any user—including a digital one—needs continued, reliable access to the network and applications. What is happening on the underlying infrastructure needs to be monitored, just like you would monitor and maintain your CRM or financial systems. There are additional considerations in relation to minor updates to platforms that interact with your automation, or background processes that update your users' desktop environment. Depending on the level of change, these could have a negative impact on some of your automated processes, so it pays to run a series of simple tests, or gradually implement the changes to your less mission-critical digital workers first.

In addition to infrastructure management, your helpdesk and technology team have to be able to receive and handle fault notifications and resolution, whether these come via the digital workforce run team elevating issues to IT or the robot itself flagging them directly.

A question we get asked a lot is this: What happens if other applications on the network with your digital worker get changed? Say the robot uses the CRM system and the CRM gets an upgrade—what then?

While some digital workforce technologies are extremely robust and will, in some cases, accommodate changes to the applications they interact with without the need for retraining, some changes can cause issues (in much the same way that they might for your human workforce).

Our advice—surprise, surprise!—is also to approach this like you would with people. You would get humans to complete user acceptance testing (UAT); this should also be done with your robots. Get clear on which applications are changing, and which digital workforce processes require testing to be done before the application is released to production. Like UAT, digital colleague acceptance testing (DCAT) is about validating that the worker can still interact robustly and correctly with the application change in order to handle all the aspects of their required work. This should be validated before the application is released into production or a change is made live. Run teams rely on technology teams to continuously communicate around application and infrastructure changes.

Fifth pillar: Lifecycle model, or digital colleague lifecycle (DCLC)

If you're familiar with the term *software development lifecycle* (SDLC), then you'll understand that the DCLC is the equivalent for your digital colleagues. The DCLC has, however, a much broader scope that includes the identification of the jobs your digital colleague can do in your business, how these are delivered, and how they are run, managed, supported, and monitored on an ongoing basis.

The final pillar in the DCAF framework concerns itself with the delivery of the full digital workforce program—from concept to retirement and everything in between. This is where the complexity of a digital workforce program becomes evident. There are so many aspects that make up this pillar—and stretch across the others—that need to be considered, decided on, and implemented.

It is also important to realize that there is no single blueprint for how to do this. So many aspects are determined by the scope and vision of your program, and the type of organization or industry you're in. For example, banks or other financial institutions are likely to require many more audit controls than some other sectors. What we present in this pillar is a typical lifecycle model (Figure 11)—one that organizations can use as a starting point and adjust to meet their specific demands.

There are a number of reasons why visualizing the process as a continuous cycle (Figure 12) makes sense. For starters, components of your digital workforce solutions should ideally be reused. The configured solutions themselves will also go

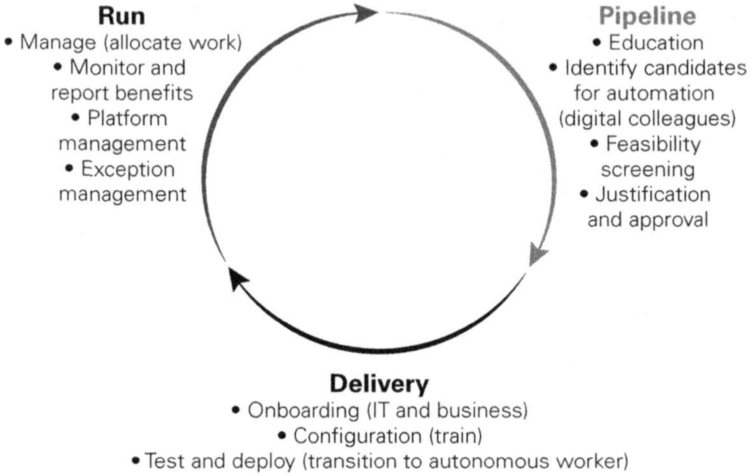

Figure 11: The Digital Colleague Lifecycle (DCLC) model

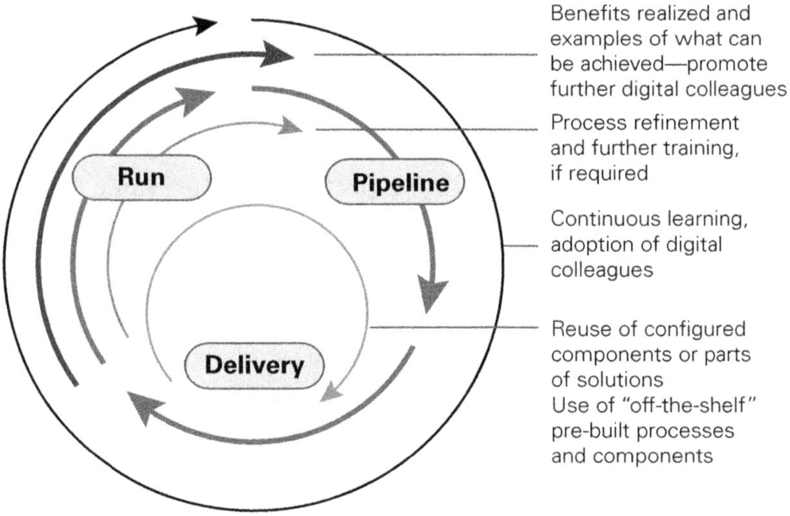

Figure 12: The Digital Colleague Lifecycle (DCLC) as a virtuous cycle

through iterative changes as either requirements change, or the systems that your solutions interact with change. They'd change in the same way that a staff member would be trained on an updated system, thereby continuously learning and evolving.

Remember the driving forces for success that we covered in Section 3? The benefits arising from digital workforce solutions, once delivered to the business, help to educate the business about what is possible for a digital workforce. The positive ROI and business benefits then spur on the demand for more digital colleagues, in a virtuous cycle. Similarly, learnings from the delivery process and from business adoption should continually help to optimize further ideation, and the delivery of more digital colleagues.

Regardless of how businesses need to adjust the various aspects that are conducted in their particular DCLC, the principle is that there are three major elements that fit together to make up a complete program. Depending on the unique shape of your digital workforce, it's likely that these could be run at a different cadence or managed differently. The run function is the continuous ongoing operation of managing your digital workforce, whereas the delivery function could operate in more of a project mode. Pipeline functions more as a continuous looping exercise as you work with different parts of your organization. All of this will be based on your strategy, as defined in the foundation of the Digital Colleague Adoption Framework.

We have included visual representations of the DCLC to further make the point that all aspects are interconnected and flow into each other. It also supports the need to expect your program to continually mature and for adoption to take time. What's important to note here is that how you set things up initially will evolve as you grow in capability and discover what works best for your organization.

Stage 1: Pipeline

Initially, we kick off the cycle at the pipeline stage (Figure 13).

As you can see from Figure 13, this phase requires multiple activities to identify and collate candidate processes, then perform assessments and validate their feasibility before approval.

Having ideation sessions with the business to collect ideas for processes is an important part of this phase, but what's key here is working closely with the business—specifically

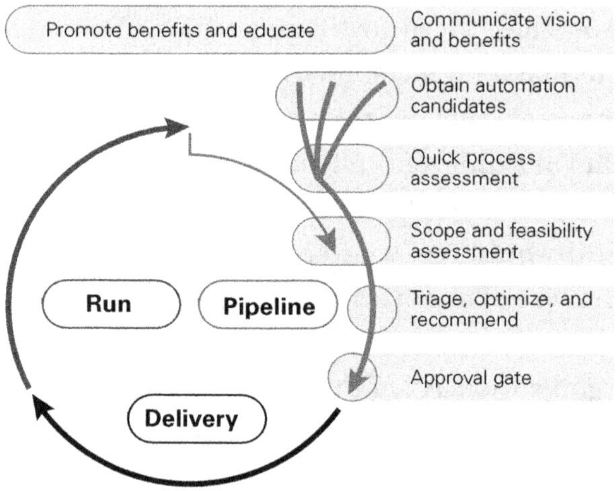

Figure 13: Kicking off the Digital Colleague Lifecycle (DCLC)—the pipeline stage

with people who are easily approachable, who understand your business's objectives and pain points, and who can educate the business managers, teams, and subject matter experts as to the benefits of a digital workforce.

It's vitally important that those within the business understand:

- what your business is aiming to achieve with automation (the vision and strategy for a digital workforce)
- what makes an ideal process or scenario for a digital workforce, and what doesn't
- how your digital workforce would interact with applications, and that it would still refer some work to users (remember, you're not targeting 100 percent straight-through processing)
- that their processes might need to be retrained should business systems change, or due to demands on the digital workforce—in other words, automation should not be sold as a silver bullet that will solve everything.

Once the business has been educated well, a large and healthy pipeline is advisable because this enables you to pick the best possible processes to automate, and provides flexibility as to the order of delivery. This provides more opportunities for efficiency—for example, doing two processes that are close to each other to enable some optimization in terms of reusable aspects of existing solutions.

It's partly a numbers game—the larger your pipeline of candidate processes submitted by knowledgeable people from within the business, the larger the number of suitable processes you will have to select from. However, what's key here, above merely the quantity of processes, is also their *quality*.

In our conversations with organizations that have successfully scaled their digital workforces and been rewarded with positive returns on their investment, many state that an abundant pipeline of candidate processes, and smart selection of which processes to automate, and in what order, are key factors in their success. They have been able to involve knowledgeable people who have not only been educated on what a good process for automation could look like, but are also on the frontline conducting these processes on a daily basis.

Conversely, we have heard from other organizations that have only been able to automate the few candidate processes that were submitted or mandated for a pilot, with very little choice. Some started by automating overly complex processes, or ones where it was virtually impossible to financially justify the expense. This is a less than ideal situation—both because it doesn't allow the automation team to learn and develop, and also because it fails to provide delivery or commercial confidence at the executive level. And we know that a lack of executive-level support is a significant barrier to success.

Again, think of collecting automation candidates like recruiting a human for your workforce. The first aspect here (education) is that you are far more likely to hire the perfect person if you can be really clear about what it is you are looking for. Once you have that, you've got more chance of getting the right person if you're choosing from a large pool of potential candidates (pipeline). Just like in people recruitment, it's critical that the selection approach is non-biased, consistent, and transparent. You will lose support, and your pipeline of recommended candidates will decline, if the business submits suggestions and never hears back about whether those processes are still up for consideration.

Smart selection of processes

Once you've got a healthy pipeline of processes—akin to a large pile of CVs—it's time to start sorting through the candidates. Ideally here you will have a way to quickly score them against some key metrics or basic criteria. To take our analogy further, you don't want to have to read every CV from cover to cover in depth; this stage is about vetting the important ones before investigating further feasibility.

Many consulting companies will encourage you to evaluate individual processes against some standard criteria that they then use to determine a score per process; then, an overall scorecard is presented to a governance board to make the final decision. While the general idea of getting a view of candidate suitability on a case-by-case basis is a fine starting point, it is far more effective to tailor the evaluation criteria to your organization's key drivers. It's common for organizations to set a simple threshold. For example, if automating a candidate process doesn't free up over two full-time-equivalent human resources to be deployed on more engaging work, then this process won't be automated. The point here is that you want a way to minimize your shortlist before performing a fuller feasibility study for each process—akin to a face-to-face interview and reference check in human recruitment.

You can't ask a process all of your most witty and engaging interview questions, sadly, but one key consideration at this reviewing stage is to evaluate how automating each process will affect other upstream or downstream processes. For example, will the automation of process B cause an increase or decrease in workload or delivery risk for the teams manually conducting processes A or C?

Lastly, one of the key benefits of RPA lies in the reusability of automation components (or objects). Therefore, it pays not to consider your processes completely in isolation. Once you have built a particular component to interact with a target application and perform a specific function, any other process being built that interacts with that same application, and needs to perform the same function, can reuse the previously built component. If this is managed well, it not only saves time and money in building new processes, but also reduces the cost and time required for process maintenance. In these ways, it can help make other processes more feasible. When evaluating candidate processes for selection, then, it is important to understand which already-built components relate to the target applications that the candidate process will interact with.

Before we look at the next phase—delivery—it's worth noting that we don't ever really leave the pipeline phase behind. Education and pipeline will be continuous activities within your organization as you develop toward a more mature digital workforce. The idea is that you're continuously generating a backlog of processes for the delivery team. We have spoken to some organizations that still have hundreds of candidate processes at various stages of their pipeline process. The Bank of New York Mellon cited this as one of their major reasons for success. But it hasn't come about by accident. According to Jon Theuerkauf, Managing Director and Group Head of Performance Excellence, BNY knew that to make the transformation successful they needed to build stakeholder buy-in and change-management practices, and get the right structures in place. They recognized that RPA needed to be treated as a business project—with a strategic vision and multidisciplinary teams working to fulfill business goals—if they were to avoid the deflection into technical issues ahead of change-management issues that too often derail IT-enabled business projects. And this thinking worked. Now, not only do they have very strong backing for the initiative, but they have also created a huge pipeline from which to select the processes with the most payback for effort, further encouraging buy-in as they deliver business value.

Stage 2: Delivery

The delivery stage of the DCLC is where a specialized team takes the automation request and translates it into a working solution for a digital colleague to run. At a high level, this includes fully understanding the detailed requirements of the process through designing, configuring, testing, and deploying. The tools and techniques used for this vary from organization to organization, and can also differ depending on the size, complexity, and criticality of the process.

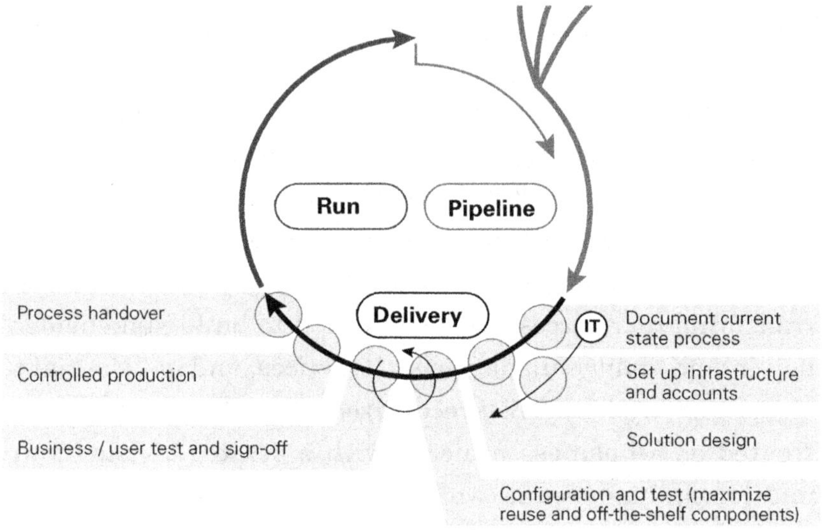

Figure 14: Building the digital workforce implementation—the delivery stage

This is the configuration stage, or the building part of the digital workforce implementation (Figure 14). It's the aspect that, perhaps unsurprisingly, a lot of other frameworks focus heavily on. However, many of these other frameworks omit a focus on reusability—the utilization of off-the-shelf components that can be employed in many cases—as opposed to building everything yourself from scratch. It is clear that this is worth investigating. However, we're not going to go too deeply into the step-by-step "how to" for delivery here—that information is relatively abundant elsewhere. Instead, we will focus more on the conceptual side of delivery.

It is also worth noting that although the DCLC captures the typical activities, it is important to engage with specialists with real-world experience in delivery. The steps taken, and the appropriate level of detail, quality, and approval during this stage, can be either too burdensome (and negatively impact delivery times and ROI) or too weak (and lead to the delivery of solutions that are unreliable, difficult to maintain and reuse, or simply don't work or meet the business's requirements).

One of the most common questions we get asked is what delivery methodology works best—Waterfall or Agile? There's no easy answer to this; we've seen examples of both working adequately. However, given the nature of delivering a robust digital workforce with its continually evolving loops, the Agile mindset seems to work well. (For more reading about Agile, head to the Resources section.) Robots, for example, may manage the main volume of standard transactions, which is then improved on—to a point—over time. Think of it, again, like training a human worker in a new role. You would teach them one element and allow them to get some proficiency in that, before adding some of the more complex elements of the job.

Also, in the same way that you would be set up for a new human joining the team, it's important to make sure that your digital colleagues are ready to show up on their first day with the right equipment and access. As soon as you have the delivery stage kicking off, make sure that you have access to all the right development environments and applications in place.

Involving the end-users during delivery

Before you start building, it is absolutely crucial to have a proper understanding of the process and validate that understanding with the people using it. In fact, regardless of methodology, you should not start building a process without fully understanding the end-to-end "happy path" (those transactions that don't require any exception management), as well as what exceptions are likely to happen. Validation by those using the process is vital for confirmation that what you are configuring is correct. It also provides the following benefits:

- It helps ensure elements haven't been left out. Even if you have the best business analysts possible documenting the processes, some important point about the process is likely to be missed. This gives you an opportunity to pick up any minor aspects that might have been overlooked.
- Working closely alongside the business, and especially with those who will be most directly affected by the digital workers, is an important part of the process to remove fear and encourage gradual acceptance.
- The opportunity to continually educate the business, build a relationship, gain trust, and set the stage for an easier adoption stage.
- When it comes to testing digital workers, users will understand what they're meant to be checking and how it all works.

As this final point implies, the relationship with the end-users isn't over once a sound understanding of the process is achieved. Don't simply take the specification away and come back a month later for a grand reveal. Keeping in close contact with the business's users is critical for routine demonstrations, validation, and testing. It pays to have experienced operators be part of the extended project team throughout the entire delivery stage, to validate that the robot is interacting with existing systems correctly, and that process happy paths are fully understood and being processed accurately.

A key point is that it is vital that the processes being delivered will trap exceptions properly and consistently, log these fully, and alert experienced operators—in much the same way that a new human user would be expected to raise concerns if they were finding unusual transactions that they didn't know how to process, or if they had issues using an application for some reason. This is to ensure the retention of the data integrity of the processes being run, which is paramount. There is a lot of emphasis on automation as improving throughput volume and speed of processes, but when you are designing and delivering processes, the highest priority should be data correctness and compliance.

Start your adoption journey early

In addition to testing, thought should be given now to how you are going to manage the change being delivered. In the pipeline stage, we talked about involving the users so that the right processes are chosen. There is another benefit—early involvement of the people who are potentially going to be affected can also help to overcome resistance to change, or fear of it. Use the delivery stage to continue with this. In our detailed look at adoption in Section 2, we commented that the level of involvement/focus here will vary depending on the level of adoption required. Whatever that level is, it's important not to forget your human users and their colleagues, and to use the techniques we've covered to help people through the change and learning curves. Investing time and effort in this area will be critical to the successful delivery of the selected processes.

Involving people in testing can also help, but don't wait until the testing stage—instead, communicate early and continuously. Utilize some of the impact assessment information that we covered earlier to better understand who to focus on and how.

Deployment: Controlled production

While different organizations address deployment in different ways, our experience has shown how effectively the "controlled production" concept works. Controlled production is where the delivery team is responsible for the new process for the first weeks of production operation, gradually increasing the volume and type of transactions conducted by the new process, and refining any tweaks required for more reliable processing and increased transactions. This also provides a gradual handover of the process from the delivery team to the run team, while ensuring that new processes are functioning correctly before being left with the business and the run team.

As you can imagine, this also helps with the reinforcing aspect of delivery. It would be a huge disappointment to the business if they got a new process automated and it only managed to process 10 percent of the transactions submitted, or didn't work every other day (much like a new hire who didn't bother turning up to work every second day would also be a bit of a disappointment!).

Stage 3: Run

The so-called "final" stage in each iteration of the lifecycle, a sound run function is critical to the success of your digital workforce.

Run is the function that monitors activities, and schedules when the digital workers need to work on different tasks (Figure 15). In some cases, the run team may be responsible for allocating more time or additional digital workforce capacity to particular processes to cope with increased demand and ensure that SLAs are met. Either way, the run team needs to fully understand the specifics of each automated process, when and how it needs to run, the required SLAs, and what to look for if it's not working right.

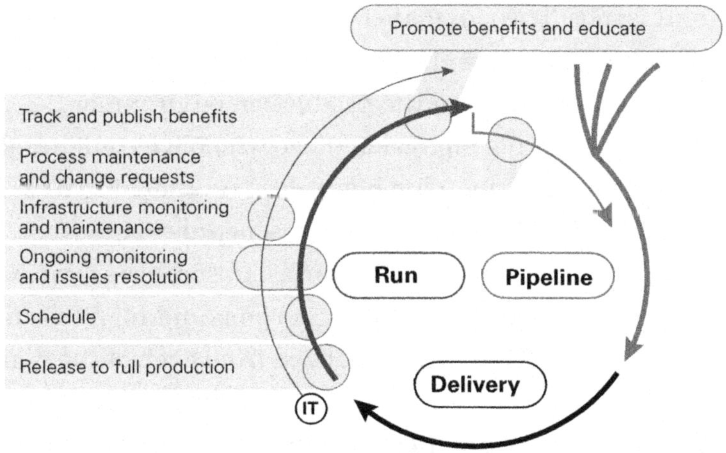

Figure 15: The final part of each iteration of the cycle—the run stage

As part of the deployment of a new process to the production environment, we expect the delivery team to prepare handover documentation and conduct a walk-through of the process with the controllers. This helps the controllers to get a sound understanding of the process, how it needs to run, and key things they will need to monitor.

At this stage, it's also the responsibility of the controller function—and run team—to ensure that any new processes being deployed to the production environment meet the required QA standards. Acting as a sort of QA gate, the run team makes sure that a digital worker isn't put into production until it is suitable for running. It's like making sure that your new human trainee is up to par before launching them onto your delicate systems. Again, this is done in conjunction with the delivery team who built the process. While a digital worker might have been QA-tested in a test environment, this can be different from a production environment, so can create different issues. The run team is responsible for ensuring that the digital worker runs in a reliable manner, by making any minor tweaks required in the production environment.

As well as working alongside those who have configured or built the robots, the run team has a close relationship with the business—to communicate status, reschedule requests, raise alerts, and respond to business queries. In some cases we have also seen the run function being responsible for recording, and reporting on, the benefits realized from each of the digital workers and presenting these in dashboards for easy viewing. In this way, they are an important part of the communication of benefits. The run team also maintains a close relationship with IT. The run team is responsible for understanding whether or not there are any system or application issues that could affect the digital workers' ability to perform their scheduled activities.

Change management continues

Just because you have deployed a process, that doesn't mean that the digital colleague change-management activities should stop. Encourage people to feed suggestions back into the pipeline and help them to feel a sense of ownership around the processes.

Exceptions will of course occur once your processes are running; consider how they are handled, and whether the team is learning from them. Have your people just become exception handlers, or do they feel empowered to make suggestions and work toward better performance and success for the whole business? The latter is the goal if you want to achieve full institutionalization. Look for areas of resistance: are there cases where people are working against a process to make it look bad? If so, keep an eye on this. True institutionalization occurs when people are committed, and adopting the processes and their new colleagues. Consider utilizing KPIs to track exception handling or, more importantly, maybe incorporate suggestions for future processes into a KPI—putting the focus on the positive rather than the negative. This will also continue to encourage a sense of ownership of both the processes and future deliverables.

A FINAL NOTE

There is a lot that goes into a digital workforce, and into working up through the phases of maturity and adoption. All the pillars of the DCAF can be done in parallel and continuously as part of any transformation program. While it's possible to automate a few processes and still see some benefit from RPA, the true potential of automation lies in *transformation* and the creation of mature internal capability to support and drive this.

CONCLUSION

The benefits provided by a digital workforce are many and various. Increased organizational agility, unrivaled efficiencies, and improved customer service are being experienced across a range of different markets by privately-held companies, government departments, and not-for-profits alike.

While plenty of organizations are implementing automation, not as many of them are doing it well—and nor are they seeing the full array of benefits. Their programs, in other words, are falling well short of their full potential. As we've described earlier in the book, research has revealed that almost half of automation programs struggle to get the positive results desired. What a waste.

We intend to change those statistics, and this book is one way, we hope, of doing so. It is our belief that with better understanding, and a more holistic approach to managing change (an approach that puts people and culture first), the failures, stalls, and significant financial wastage that we're now seeing in the industry can be avoided.

A digital workforce might, at first glance, appear to be a technology project, but it really is so much more than that. People often kick off the venture thinking that it's simple, and because it looks like technology they give it to the

IT people or to IT companies to implement. Sure, putting in some automation will generate some benefits, from a strict technology-implementation point of view—like faster processing, 24/7 operations, and stacks of resources and FTEs saved. However, organizations that take only this approach are missing the opportunity to utilize a digital workforce to create lasting change. They're missing the opportunity to truly transform.

> True transformation is driven from the people in your organization.

When done well, the implementation of a digital workforce represents a significant cultural shift for your organization, along with benefits for the people in it and, hopefully, society. The difference comes down to different mindsets around adoption, and the level of maturity (in terms of capability) that your organization aspires to for your digital workforce. In other words, the business-people need to do a big part of the thinking up front.

If you're dipping a toe into automation—what we refer to as the *initialization* stage—then it's true that you'll achieve some gains. But it's still mainly about potential. You're only getting started.

If you're looking to scale your digital workforce throughout your organization—reaching *industrialization*—then you'll be looking for greater engagement from your people. You'll be working to demonstrate and communicate the benefits more widely, gain efficiencies in delivery, and build a healthy pipeline across the business. Your team will be starting to see the value in your digital workforce.

Finally, if you want to reach the level of *institutionalization* or *true transformation*—and achieve the full potential of a digital workforce—you can only do this if your program is owned and driven by the business. A transformative digital workforce is led by the business, and should—by necessity—be driven from the strategic goals of your organization. That's because effective and lasting change comes from a holistic view that masters all of the individual, but interconnected, aspects of automation. And your team recognizes it.

> Adopting a digital workforce is an organization-led initiative. Only when people are brought on the journey will you reap the real benefits.

As well as needing a business, rather than a technology, focus, a transformative digital workforce puts a change-management focus ahead of (or much more broadly than) a linear project focus. Undoubtedly the technology that underpins automation is important, but a digital workforce has always been far more about people than about technology. And anytime you are effecting a significant cultural shift in people, you need to proactively and positively manage that change. This remains the aspect that is often overlooked, despite being the key part of getting any digital workforce implementation right.

For any successful automation project the aim should be the creation of an *augmented* or *unified* workforce. And that's what a digital workforce means, by definition—it's the technologies that work in collaboration, both with each other and with your human workforce, to undertake key business functions. So, the intention is that your digital workforce works alongside your existing human workforce, and that both are established, educated, trained, and managed in a way that encourages them to work together.

Key to this success is maximum adoption, and key to maximum adoption is the utilization of a robust and holistic framework; hence the creation of our Digital Colleague Adoption Framework (DCAF). Once the leadership within the business has identified the "why," they are better able to provide the conditions that the project team needs to execute the "how."

The specifics surrounding the "how" will vary with each implementation and will change with time, especially in a field that is evolving as rapidly as this one is. We're seeing seismic shifts in the way people work; the impact that these changes have on organizations' workforce operating models is a real driver for doing this well.

The good news is that the strategic and conceptual principles in this book will remain valid. They represent sound management, and those practices will endure even as the technology and, to an extent, the way we work shifts. The core of this thinking has already evolved over all of the months we've spent writing this book.

Together we can change the course of how people work, optimize organizational activities, let people do what they

do best, and create a better society and world. It's called revolutionizing work—let's do it.

We intend this book to become a "living" entity. We are already working to extend it into a much more prescriptive framework, with all the aspects and methodologies for successful implementation covered. We have also established a corresponding website that will enable us to facilitate a real-time forum of ideas, and share more resources as they are developed.

We are committed to continuing to evolve with the industry, and hope that you will join us there.

WANT TO KNOW MORE?

Join us at www.DCAF-book.org

GLOSSARY

Term	Description
adoption	Related to change management, adoption refers to the level of uptake of a particular solution or change.
Agile (delivery)	A methodology initially created as a solution to problems existing in traditional (*see* Waterfall) approaches to software development, which has evolved into a way of delivering technical and non-technical solutions. Agile methodology involves embracing change, close collaboration with the customer, delaying detail until needed, and incremental delivery. Common examples of Agile methodologies include SCRUM, the Dynamic Systems Development Method (DSDM), and the Scaled Agile Framework® (SAFe)—more about which can be found in the Resources section of this book.
application	A piece of computer software built to provide a specific set of functionality for an end-user (an end-user could be a human user, a digital colleague, or another application).
application programming interface (API)	A software component that allows two or more applications to talk to each other in a well-defined and controlled manner. It defines the way in which you can talk to a particular software application, and the protocols used.

Term	Description
artificial intelligence (AI)	A generic description for a "smart/thinking" computer system or application that is built to undertake tasks that would normally require a level of human intelligence. *See also* machine learning
augmented workforce	The combination of digital and human workers/ colleagues.
automated process	A process containing one or more steps to achieve a well-defined outcome that has been automated.
automated workforce	*See* digital workforce
automation	The act of automating a primarily human interaction or process. The removal of one or more human interactions to complete a set of tasks/process steps.
automation center of excellence	A team that provides leadership, training, and best practice for automation.
big data	An extremely large dataset that requires computational analysis to create information and knowledge. Often used in predictive and trend analysis.
business process management (BPM)	Historically tied with workflow management, this is a technology that pre-dates RPA and is used to automate, track, and manage workflow for business processes.
change management	A set of tools, techniques, and methodologies that can be used to help a human workforce through the changes resulting from transformation, while enabling the required level of adoption of the new digital workforce.
compliance	The first level of adoption of a digital workforce by its human colleagues.

Term	Description
Connected RPA	*See* digital workforce
customer relationship management (CRM)	Software for management of customers and prospects.
design authority	A role in the digital workforce delivery team that provides guidance across multiple aspects of the DCLC, including candidate process selection, solution designs, delivery standards, and quality assurance framework.
digital colleague	A robot that handles one or more process automations to achieve beneficial outcomes. Comparable to a human worker.
digital colleague acceptance test (DCAT)	A user acceptance test for your robots, testing that the robotic process/digital colleague undertakes the work in the way that is expected by the business owners.
Digital Colleague Adoption Framework (DCAF)	A framework to ensure business ownership and the setting of required business conditions to assist the successful adoption of a digital workforce in an organization.
Digital Colleague Lifecycle (DCLC)	An adaptable starting framework to help organizations structure their unique digital workforce implementation program to their specific needs and business conditions.
digital workforce	A collection of technologies that work together to undertake automated tasks. A digital workforce works alongside humans, taking away the mundane or repetitive aspects of their work.
downstream processes	Latter processes that will subsequently be affected by the automation of the current process under investigation.

Term	Description
enterprise resource planning (ERP)	In the technology sense, an integrated system, or set of systems, that manages key parts of the business, such as finance, HR, supply chain.
exception management	A process defined to manage the handling of unplanned outputs from an automated process through to resolution.
governance	The controls put in place to ensure that IT does what it is supposed to do, in a way that is beneficial to the organization. Often includes resource management and control of the investment approval processes.
happy path	Often used to define the execution of a process in a manner that should not result in any errors or exceptions.
information technology (IT)	The set of systems for storing, retrieving, processing, and sending information.
infrastructure	The set of hardware, software, network, and related equipment upon which the digital workforce is built and deployed.
intellectual property (IP)	Copyright, patents, trademarks and trade secrets developed by businesses.
intelligent automation	A combination of RPA and artificial intelligence to enhance the results and automation of a business process.
machine learning	A subset of artificial intelligence that deals with the ability for computer systems to learn and adapt automatically through experience.
maturity	The level of maturity that an organization has toward automation, comprising three levels: initialization, industrialization, and institutionalization.
optical character recognition (OCR)	**A technology that reads scanned documents or electronic files.**

Term	Description
organizational workforce	An augmented workforce, using both human and digital workforces.
Pareto principle	Also known as the 80/20 rule, this is a principle suggested by Joseph M. Juran stating that for many events roughly 80 percent of the effects come from 20 percent of the causes.
pilot	Releasing the robotic process in a controlled, limited-scope environment to test the outputs/outcomes of the automated processes.
quality assurance (QA)	Providing confidence to the business owners/stakeholders that the quality of an automated process meets the required level (is fit for purpose).
return on investment (ROI)	A performance measure used to validate an investment or compare a set of investments, by calculating the value of the return on a particular investment against its cost.
robot	*See also* digital colleague
robotic operating environment	An infrastructure stack set up to manage and execute a set of digital colleagues/software robots.
robotic process automation (RPA)	A form of automation of business processes built on a specialized software platform. Often called bots/robots.
satellite application	*See* application
service level agreement (SLA)	The definition of the level of service required by one party from another party (e.g. a customer from their supplier). Usually defined in measurable and contractual terms.
software development lifecycle (SDLC)	A process used by the software industry to progress from a set of requirements to an implemented solution in a phased, sequential approach.

Term	Description
standard operating procedure (SOP)	A set of documented steps to perform routine, often complex tasks.
straight-through processing	A method used to quantify the number of transactions processed without exception or the need for manual intervention. Usually presented as a percentage within a given period.
subject matter expert	An authority in a particular topic. In the context of digital automation this refers to the business representative who fully understands the business process and its value to the organization.
technology stack	The set of technologies that an organization uses to deliver a digital workforce.
unified workforce	The combination of a human and a digital workforce to deliver a set of organizational outcomes. *See also* organizational workforce
upskilling	The training/learning required by the human workforce to best utilize the transformation into a unified workforce and the adoption of a digital workforce.
upstream processes	Processes that can have an effect on the process currently being automated, or being considered for automation.
user acceptance testing (UAT)	Testing an activity to ensure it meets business, not just technical, requirements.
Waterfall	A software development lifecycle model that delivers the phases of software development (often requirement, design, development, test, implementation) in a sequential fashion.

RESOURCES

Additional resources

Digital Colleague Lifecycle (DCLC)

We covered the details of the DCLC and the impact of the foundation and the five pillars on each element of the lifecycle in the main body of the book; however, we have found it useful to apply a mindset related to the "why" of the lifecycle

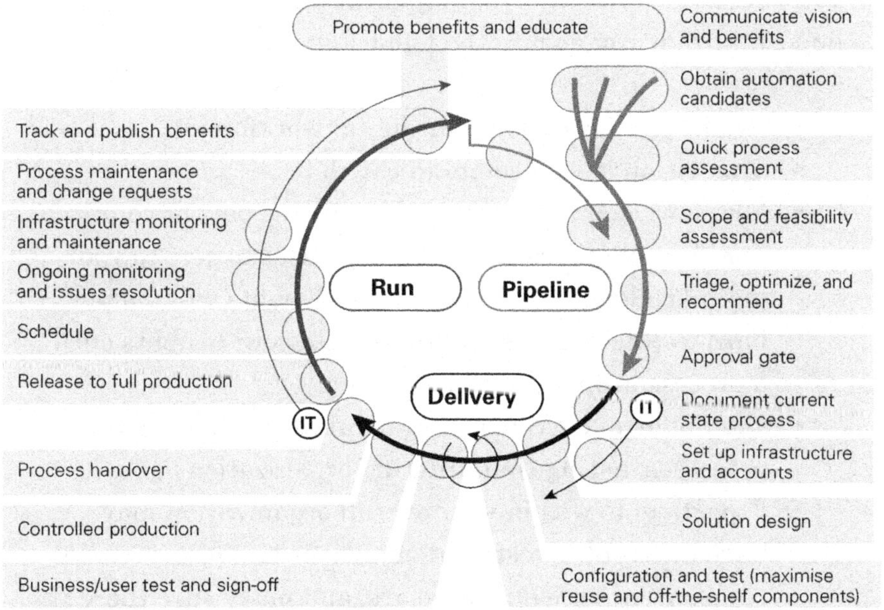

Figure 16: The Digital Colleague Lifecycle (DCLC) at its best—operating as a virtuous cycle

when planning how to implement it within your organization.

First, it should be noted that the DCLC is truly cyclical if implemented successfully (there are some questions/tips later in this section that can help achieve success here). This cyclical nature means that although you *can* loop through this lifecycle once and stop, by doing so your organization may miss out on truly leveraging the benefits of your digital colleagues. Promoting the benefits and increasing education and awareness of the digital colleagues can lead to a virtuous cycle of increased demand into the pipeline.

Pipeline

For successful pipeline implementation we have found it important that you ensure the following:

- You have a vision for your digital workforce.
- This vision is well communicated.
- You have clear, measurable benefits as part of your justification/approval of new processes to give to your digital colleagues. A lack of clear benefits will make it hard to track, publish, and promote your benefits once a process is operational.
- You undertake a consistent justification for selected processes, noting that differing organizations and departments within your overall organization may have different investment objectives and priorities. Be transparent on what these are, and make sure there is consistency to your decision-making.

- You have a change-management strategy that can be continued in the future phases of the lifecycle. It is important that aspects such as communications, planning for resistance, and adoption are put in place for individual processes and at a governance level across the lifecycle.
- You have a clear strategy for governance of your digital workforce. This should include not only how approvals will be undertaken to ensure consistency, but also things like how are you managing infrastructure investment, operational ownership, and realization of benefits.

Delivery

This phase is focused on the delivery of the outputs—the processes that your digital colleagues are going to be running in operations. Even though this phase can be focused on more traditional project delivery methodologies, it is important not to overlook the need to incorporate change-management techniques throughout this stage in parallel to, and at the same level of importance as, project delivery.

For a successful delivery implementation we have found it important that you:

- Involve the business in solution design and testing.
- Ensure that there is clear business ownership of the process.
- Ensure that key infrastructure requirements (around technology aspects such as user access to

infrastructure, security, etc.) are in place or planned as part of the project. These are often overlooked and cause problems, especially in early instances of process creation.

- Ensure that you understand the different levels of diligence required for different processes—dealing with mission-critical processes vs. more back-office, mundane processes.

Run

This phase is focused on the operational aspects of the digital workforce—the "business as usual" running of the process. As such, it can be easily forgotten about. It is important to keep the processes and their operations visible, even at a senior level within your organization. Generating a virtuous cycle for further improvements and benefits to this and other processes will require ongoing monitoring, changes, and tracking of benefits. This phase generates the outcomes that lead to the realization of benefits initially used to justify process automation.

For a successful run implementation we have found it important that you ensure:

- Regular monitoring of the processes.
- Regular monitoring of exceptions generated (these could lead to change requests in the pipeline phase of the lifecycle for the realization of further benefits).
- There is clear ownership of the operational digital colleagues. This could be centralized or decentralized.

- There is clear ownership of the infrastructure upon which the digital colleagues are deployed.

Some useful tools and techniques

Change management

As you will have noticed, throughout this book we have made reference to specific tools and techniques that can be used to aid the thinking around the change impacts of automation. There is also a heavy reliance on change management throughout the Digital Colleague Adoption Framework (DCAF).

While we have referenced some of the great thinkers in terms of change management, it is not our intention to provide in this book those tools and techniques that can be applied to your automation journey. Our website (www.DCAF-book.org) is the "living" document for that sort of thing. However, some of the learnings that have been made over the years shouldn't be forgotten—they are as relevant today as when they were first proposed. Accordingly, we have included some background on specific thought leaders and useful articles in the change management space.

We have used the McKinsey 7S model as a linchpin to understanding the integrated aspect of organizations, and the interrelationships and impacts that aspects of an organization can have on each other. The McKinsey model was first proposed in the 1970s and later incorporated in the book *In Search of Excellence* by Tom Peters and Robert H. Waterman.

For a complete source of change-management tools and techniques (of which there are a huge number), there are two primary texts we recommend:

- *The Effective Change Manager: The Change Management Body of Knowledge,* published by the Change Management Institute (https://www.change-management-institute.com/cmbok)
- *The Effective Change Manager's Handbook*, edited by Richard Smith, David King, Ranjit Sidhu, and Dan Skelsey, and published by Kogan Page, 2014.

Both of these texts cover a comprehensive list of change-management tools and techniques that are applicable to an automation journey. In our current world, which is driven by the need to be adaptable, and where change seems to be a constant, keeping up with the pace from a change management perspective is becoming harder and harder. A new variation on traditional change-management viewpoints is needed, and these are indeed appearing. One of the authors of *The Effective Change Manager's Handbook* (Dan Skelsey) provides insights and coaching into how this can be achieved through Rapid Agile Change Management. Links to this can be found on our website (www.DCAF-book.org).

Agile delivery/project management

This book is not a guide to Agile delivery, of course, but it should be noted that Agile is an ever-growing technique for delivering solutions, and can easily be applied to the delivery stages of the DCLC. With this in mind, we thought it would be useful to include some resources that you can review to investigate this methodology further.

Agile techniques are founded around a manifesto that can

be found at https://agilemanifesto.org. The Agile manifesto deals with the development of software (and can easily be used for other technical solutions). The Agile manifesto is as follows:

> We are uncovering better ways of developing software by doing it and helping others do it. Through this work we have come to value:
> **individuals and interactions** over processes and tools
> **working software** over comprehensive documentation
> **customer collaboration** over contract negotiation
> **responding to change** over following a plan
> That is, while there is value in the items on the right, we value the items on the left more.

We find that applying the manifesto mindset can aid in the successful delivery of automated processes. For example, while it doesn't preclude the need for a solution design, it does recommend that you do enough design work up front to understand what you are trying to deliver.

There are numerous Agile techniques and frameworks around. Some of the more common ones are:

- Agile Business Consortium: https://www.agilebusiness.org
- SAFe: https://www.scaledagileframework.com
- Scrum: https://www.scrum.org

We are not going to go into the pros and cons of these in the context of their ability to deliver—they all have very useful techniques to apply depending on your unique set of circumstances and requirements. However, we will say that if you choose to utilize one of these techniques to aid in your delivery, you should ensure that you employ (internally or externally) practitioners who are highly skilled in the particular methodology.

Off-the-shelf components

As we have mentioned, off-the-shelf digital workforce components help to improve both the ROI and robustness of the process being automated. Below are some key resources for obtaining free and "available to purchase" reusable components:

- www.RPA-marketplace.com
- www.blueprism.com/products/digital-exchange
- https://connect.uipath.com/marketplace

For the most up-to-date information about what is happening in the ever-evolving digital workforce domain be sure to check out:

- www.DCAF-book.org
- www.q4associates.biz
- www.meetup.com (for your local meet-up for RPA and AI)

NOTES

1. James Manyika, Susan Lund, Michael Chui, Jacques Bughin, Johnathan Woetzel, Parul Batra, Ryan Ko, and Saurabh Sanghvi, "Jobs lost, jobs gained: What the future of work will mean for jobs, skills, and wages," McKinsey Global Institute, November 28, 2017. https://www.mckinsey.com/featured-insights/future-of-work/jobs-lost-jobs-gained-what-the-future-of-work-will-mean-for-jobs-skills-and-wages
2. Leslie Willcocks, in "The value of robotic process automation," interview with Xavier Lhuer, March 1, 2017, McKinsey & Company, Financial Services, Our Insights. https://www.mckinsey.com/industries/financial-services/our-insights/the-value-of-robotic-process-automation
3. Leslie P. Willcocks and Mary C. Lacity, "OWS18 findings revealed on client service automation deployments—what do they mean for your job and organization?", *Pulse Magazine* (published by IAOP). https://iaoppulse.net/research-corner-ows18-findings-revealed-on-client-service-automation-deployments-what-do-they-mean-for-your-job-and-organization/
4. Jared Wadley, "Word choice: Hidden meanings can influence our judgment," *Michigan News*, University of Michigan, June 6, 2016. https://news.umich.edu/word-choice-hidden-meanings-can-influence-our-judgment/
5. Michael Levine, "Logic and emotion. Delving into the logical and emotional sides of the human brain," *Psychology Today,* July 12, 2012. https://www.psychologytoday.com/us/blog/the-divided-mind/201207/logic-and-emotion

6. John Narver, Stanley Slater, and Brian Tietje, "Creating a market orientation," *Journal of Market-Focused Management*, 1988, vol. 2, pp. 241–255. doi: 10.1023/A:1009703717144
7. Lars Faeste, Jim Hemerling, Perry Keenan, and Martin Reeves, "Transformation: The imperative to change," Boston Consulting Group, November 3, 2014. https://www.bcg.com/publications/2014/people-organization-transformation-imperative-change
8. Gina Barlow and Andrew Tubb, "Driving business performance—Project Management Survey 2017," KPMG New Zealand, 2017. https://home.kpmg/content/dam/kpmg/nz/pdf/01816%20PMS%20ReportCompressed.pdf
9. Tony Schwartz, "Leaders focus too much on changing policies, and not enough on changing minds," *Harvard Business Review*, June 25, 2018. https://hbr.org/2018/06/leaders-focus-too-much-on-changing-policies-and-not-enough-on-changing-minds
10. Tom Peters and Robert H. Waterman Jr., *In Search of Excellence*, Harper & Row, 1982.
11. Bryan Walker and Sarah A. Soule, "Changing company culture requires a movement, not a mandate," *Harvard Business Review*, June 20, 2017. https://hbr.org/2017/06/changing-company-culture-requires-a-movement-not-a-mandate
12. Ibid.
13. Maria Juncu, "5 ways to lead in an era of constant change," *Performance Magazine,* April 29, 2017. https://www.performancemagazine.org/lead-era-constant-change/
14. John P. Kotter, *Accelerate: Building strategic agility for a faster-moving world*, Harvard Business Review Press, 2014.
15. Tracy Brower, "Successful change management: 6 surprising reasons people resist change and how to motivate them to embrace it instead," *Forbes,* February 16, 2020.
16. Rosabeth Moss Kanter, "Ten reasons people resist change," *Harvard Business Review*, September 25, 2012. https://hbr.org/2012/09/ten-reasons-people-resist-chang

17. Kelman, H.C. "Compliance, identification, and internalization: Three processes of attitude change," *Journal of Conflict Resolution*, 1958; 2(1): 51–60.
18. Javier Gil Gómez, Pablo Hernández, and Rafael Ocejo, "Four success factors for workforce automation," McKinsey & Company, May 28, 2019. https://www.mckinsey.com/business-functions/operations/our-insights/four-success-factors-for-workforce-automation#
19. Richard Smith, David King, Ranjit Sidhu, and Dan Skelsey (eds.), *The Effective Change Manager's Handbook: Essential guidance to the change management body of knowledge*, Kogan Page, 2014.
20. Daniel H. Pink, *Drive: The surprising truth about what motivates us*, Riverhead Books, 2009.
21. Ibid.
22. Ibid.
23. Gómez et al., "Four success factors for workforce automation"
24. George Lawton, "Hyperautomation: The most significant RPA trend of 2020," *SearchCIO*, February 28, 2020. https://searchcio.techtarget.com/feature/Hyperautomation-The-most-significant-RPA-trend-of-2020
25. Ibid.
26. Everett M. Rogers, *Diffusions of Innovation*, Simon & Schuster, 1962.
27. Gómez et al., "Four success factors for workforce automation"
28. Rogers, *Diffusions of Innovation*
29. Lawton, "Hyperautomation"
30. Gómez et al., "Four success factors for workforce automation"
31. Lawton, "Hyperautomation"

ACKNOWLEDGMENTS

As with everything in this book, we researched this section too. One author commented that this was one of the hardest parts of the book to write—and we have to agree. Support, input and thinking has come from a broad range of people, spanning different sectors and geographical areas over a long period of time, and has included very practical insights, trial and error learning, and some amazing research from an academic perspective. Obviously, none of this would have been possible without the continued support, contribution and sacrifices of our spouses, Denise and Irina. You know you are still loved even when we may be sitting there in person, but far away thinking of robots!

Next we need to acknowledge the hundreds of people from around the world with whom we have discussed the successes and challenges of adopting a digital workforce within the day-to-day business environment, at conferences, and online. The following people have stood out in regards to leading the thinking, trialing, questioning, and building a body of knowledge in this domain: Shayde Boryer and Sanjay Giridharan at Q4 Associates; Barry Coughlan and Rachel Swindells of ASB Bank; Craig Richardson of MMC Funds Management; Abhi Kadian of Westpac Group; Vincent Ma at IBM; and the incredibly knowledgeable teams from Blue Prism and UI Path.

Lastly, but no less important, are the amazingly talented teams who helped with the creation, review, and critique of this book. These include Christina Wedgwood and the supportive and knowledgeable team from Intelligent Ink, and Debra Millar from Point Publishing. We joke about us supplying the ingredients and these guys baking the dish—one so delicious and tasty that people can't put it down! Hopefully this is what we have achieved.

We love the saying that "no two people ever read the same book", so please take out what is most relevant to you—and feel free to leave the Brussels sprouts. Bon appetite!

Maurice and John

Made in the USA
Monee, IL
23 April 2021

66564983R00125